PROGRESS IN FOOD SCIENCE AND TECHNOLOGY

LACTIC ACID

PRODUCTION, PROPERTIES AND HEALTH EFFECTS

PROGRESS IN FOOD SCIENCE AND TECHNOLOGY

Additional books in this series can be found on Nova's website under the Series tab.

Additional e-books in this series can be found on Nova's website under the e-book tab.

BIOCHEMISTRY RESEARCH TRENDS

Additional books in this series can be found on Nova's website under the Series tab.

Additional e-books in this series can be found on Nova's website under the e-book tab.

PROGRESS IN FOOD SCIENCE AND TECHNOLOGY

LACTIC ACID

PRODUCTION, PROPERTIES AND HEALTH EFFECTS

ALFONSO JIMÉNEZ MIGALLÓN
AND
ROXANA A. RUSECKAITE
EDITORS

Nova Science Publishers, Inc.
New York

For permission to use material from this book please contact us:
Telephone 631-231-7269; Fax 631-231-8175
Web Site: http://www.novapublishers.com

Library of Congress Cataloging-in-Publication Data

Library of Congress Control Number: 2012940134
ISBN: 978-1-62081-643-1

Published by Nova Science Publishers, Inc. † New York

Contents

PREFACE

The development of new materials of natural origin and biodegradable characteristics for many applications is one of the most important challenges for the materials research groups and polymer industry during the next few years. Poly(lactic acid), PLA, is a biodegradable polymer synthesized from lactic acid (LA), which on its side is obtained from renewable resources. The editors have carried out a selection and seemingly have chosen enough of good reviews and unique works in the branch of lactic acid and its properties, polymerization and final applications of PLA. Apparently, these works are at the front edge of science and will be of interest for high readership of scientists and students in the branch of chemistry and physics of polymers and biopolymers. The editors expect that the readers will be happy with this selection.

Chapter 1 - Lactic acid can be manufactured either by chemical synthesis or by fermentative processes. In recent years, the amount of lactic acid obtained by biotechnological methods has increased. The highest cost in the traditional process of lactic acid production by carbohydrate fermentation lies in the separation steps that are needed to recover and purify the product from the fermentation broth. To reduce these costs, many studies on lactic acid separation have been conducted using different separation techniques. In this chapter the purification of lactic acid by means of membrane techniques and ion exchange are considered. The separation efficiency of nanofiltration membranes in the treatment of ionic solutions is explained by the combination of size and charge effects. A strong influence of pH on lactic acid transport through nanofiltration membranes is observed. The pH value affected both rejection and permeate flux. Rejection increased with pH while flux decreased with this variable. At the acidic pH required to convert lactate into undissociated lactic acid (pH 3.0), the recovery of lactic acid is low (around 50 %) and similar to that of the inorganic salts present in the fermentation broth. Therefore it should be improved in order to reduce lactic acid losses and increase purity. Special interest has received the recovery of lactic acid using bipolar membrane electrodialysis due to the possibility of obtaining lactic acid and caustic soda simultaneously from pretreated fermentation broths. In the second part of the chapter a review on different processes for lactic acid recovery based on ion exchange is performed. Most of the works are focused on the recovery of lactate ions from the broth by means of anion exchangers However, the separation process proposed in this chapter is an ion-exchange process that makes it possible to remove undesired ions from the fermentation broth, while lactic acid remains in solution. The final purity of lactic acid was higher than 99%. Losses were estimated to be lower than 15 %.

Chapter 2 - Lactic acid has been used in pharmaceutical, cosmetic, chemical and food industries. It can be used as a raw material in biodegradable plastics, such as polylactic acid (PLA) as well as to improve the physical properties in the production of garbage bags, agricultural plastic sheeting, and food packaging. It can also be used in sutures and surgical implants due to its biocompatible and bioabsorbable characteristics. Lactic acid is industrially produced either through chemical synthesis or microbial fermentation.

The advantage of the biological method is that an optically pure lactic acid can be obtained by choosing a strain of lactic acid bacteria, whereas chemical synthesis always results in a racemic mixture of lactic acid. The optical purity of lactic acid is very important to the physical properties of PLA and obtaining a more stable crystalline polymer than that achieved with a racemic lactic acid. Therefore, raw materials for the industrial production of lactic acid need to have characteristics such as low cost, low levels of contaminants, rapid fermentation, and year-round availability. The cost of raw materials represents 68% of the total cost of lactic acid production. A number of industrial byproducts or wastes have been evaluated as substrates for lactic acid production with the aim of decreasing the cost of the process, such as sugarcane, molasses, cassava wastewater, and whey as carbon sources and corn steep liquor (CSL) and yeast autolysate as nitrogen sources. In this chapter, it will be discussed the fermentation technology to lactic acid production that has been used and exploited by using the cheaply available source materials.

Chapter 3 - Siloxane-containing vaterite (SiV)/poly (L-lactic acid) (PLLA) hybrid fibremats (SiPVHs) have outstanding potential for bone regeneration applications. SiPVHs were prepared by using melt-blending and electrospinning protocols and modified by increasing PLLA content from 40 to 70 wt% (denoted by SiPVH60 and SiPVH30, respectively). The tensile strength of SiPVH30 was 10 times higher than that of SiPVH60. The elongation before the failure of SiPVH30 was twice as large as that for SiPVH60. The necking of fibers occurred at the failure points in SiPVH30 but not in SiPVH60. Ionic silicon-species in Tris-buffer solution from SiPVH30 was the relatively constant release. SiPVH30 could also be coated with hydroxyapatite by being soaked in simulated body fluid. A new type of membrane was prepared by bonding SiPVH60 fibremat with PLLA. The bilayered membrane showed flexibility, relatively high mechanical strength to allow easy handling, and not only functions as the barrier to control the intrusion of soft tissue but also the function of a 3-D scaffold which enhances the cell activation and the bone formation.

Chapter 4 - Lactic acid (2-hydroxypropanoic acid) is one of the smallest optically active molecules which can be either of L(+) or D(-) stereoisomer. This is produced by animals, plants, microorganisms in the nature and can also be derived from intermediates with an origin in renewable materials or from chemicals derived from coal or oil. The research on lactic acid based polymers [poly(lactic acid)-PLA-] intended for medical applications has been accelerated since 1960 and during the two last decades an increasing activity on large scale industrial lactic acid based polymers for environmental uses has been also noticed, probably due to the recent need for the urgent development of biodegradable materials that would not involve the use of toxic components in their manufacture and could be degraded in relatively short times to environmentally friendly products.

The main limitations of this polymer towards its wider industrial application are its poor thermal and mechanical resistance and limited fire retardance properties, restricting its access to industrial sectors, such as transport applications, in which its use would be justified when biodegradability is required. Nevertheless, it has been found that the above drawbacks can be overcome by enhancing their thermal and mechanical properties through copolymerization, blending and filling techniques. Indeed, the addition of nano-sized fillers could potentially confer multifunctional enabling properties to this polymer. In this way, nanoparticles in biodegradable polymers have a strong promise in designing eco-friendly nanocomposites for several applications. A fairly new area of composites has emerged in which the reinforcing material has the dimensions in nanometric scale, such composites are called nanocomposites. It has been proved that these nanocomposites can significantly improve thermal and mechanical properties of several polymers even with a very low level of nanoparticle addition due to their nanoscale dispersion, so that the reinforcement efficiency of nanofillers corresponds to that of 40-50% loading of classical micro-fillers.

This chapter was addressed to review the effect of addition and co-addition of different nanoparticles on the thermal, mechanical and fire retardant properties of PLA. As general result, potential bio-based compositions for industrial applications were found with the mixing of PLA and these used nanoparticles, achieving important enhancements in the rigidity and in fire retardance of the PLA matrix. Even more, it was found that the co-addition of thenanoparticles in the polymer matrix resulted in the occurrence of synergistic interactions between nanoparticles on the above properties, therefore regarding these compositions as potentially interesting materials for the enhancement of PLA properties for transport applications, electric and electronic engineering materials.

Chapter 5 - The industry of minimally processed vegetables is constantly searching for new technologies for maintaining quality and inhibiting undesired microbial growth in all steps of production and distribution. New biopreservation technologies are recently considered as a promising perspective in order to reduce the chemical preservatives. The aim of this study was to determine the influence of added bacteriocin-producing lactic acid bacteria (LAB) on native microbial population associated to lettuce. The authors also determined the "*in vivo*" protective effect of LAB against *E.coli O157:H7*. Lettuces were inoculated with single LAB strains (*Enterococcus faecium, Lactococcus lactis, Enterococcus hirae* and *Enterococcus canis*) or in combination with a pathogenic strain of *E.coli O157:H7*. The evolution of microorganisms in lettuce was followed during storage at 8 °C. Although the treatments did not completely eliminate the pathogen from lettuces, the results showed that LAB had a good colonization in both substrates. Biopreservation effect exerted by LAB on the *E.coli O157:H7* population was bacteriostatic for the case of *E.faecium* and *E.hirae*. Microbial competition strategies require time, which in products such as lettuce could produce the rapid quality deterioration. Therefore, it would be of great interest the study of these strains in products having a longer shelf life.

Chapter 6 - Lactic acid is a very important chemical compound for food, chemical and pharmaceutical industries. Regarding the food industry applications of lactic acid, it is generally used as natural preservative and as an acidulant. It is usually produced through fermentation of various raw materials. However, it is estimated that approximately 90% of the total production cost of lactic acid is due to these raw materials. Likewise, the research community makes efforts to develop new technologies and to exploit attractive agro-industrial by-products targeting to the decrease of the whole production cost. In this pattern, corn sugars, molasses and cheese whey have been extensively examined. However, cheese whey seems to offer the more preferred raw material, since it is cheap, abundant and responsible for environmental pollution due to its high organic load. Several studies have been published the last decade, suggesting ways for the exploitation of cheese whey. On the other hand, the appropriate choice of the bacteria that will degrade cheese whey is of great importance. For this purpose several lactic and bacteria (LAB) have been applied and quite encouraging results have been emerged. Among LAB *Lactobacillus helveticus* is the most effective, due to the achievement of higher lactic acid productivities compared to other LAB. In addition application of *Streptococcus* has also led to satisfying results. Finally, genetic modification for the development of a satisfactory and efficient LAB strains for cheese whey fermentation conduction has also been studied, as well as various immobilization techniques that have also been successfully applied. In the frame of this review the importance of lactic acid and the most significant research efforts for enhancing lactic acid production are discussed.

Chapter 7 - This chapter reports the relevance and the promising perspectives of poly(lactic acid) (PLA) based nanostructured biomaterials in tissue engineering. The development of PLA matrix nanocomposites with different morphologies and the effects of their surface nanotopography are here extensively investigated in terms of polymer physicochemical properties and stem cell response. Aliphatic polyester polymers, such as PLA, have already attracted wide attention for the preparation of biomaterials taking advantage of their biodegradability and biocompatibility in the human body. Polymeric nanocomposite scaffolds represent a valuable strategy that offers the opportunity to modulate the surface, mechanical and chemical properties of polymer by adding a specific type and content of nanostructure to the biomaterial. The possibility to control cell function by modulating the polymer properties represents a key point of material science in tissue engineering applications. Combining surface research efforts from the physical and chemical approaches may lead to a wide range of cellular responses and activities. Nanostructured biodegradable polymeric materials are ready for take off and certainly promise an exciting future at the interface of chemistry, biology and material science.

In: Lactic Acid
Editors: A. Migallón and R. Ruseckaite

ISBN: 978-1-62081-643-1

Chapter 1

PURIFICATION OF LACTIC ACID OBTAINED BY FERMENTATIVE PROCESSES BY MEANS OF MEMBRANE TECHNIQUES AND ION EXCHANGE

***F. A. Riera*[1] *and S. Alvarez*[2]**

[1]Department of Chemical Engineering and Environmental Technology, University of Oviedo, Oviedo, Spain

[2]Department of Chemical and Nuclear Engineering, Universidad Politécnica de Valencia, Valencia, Spain

ABSTRACT

Lactic acid can be manufactured either by chemical synthesis or by fermentative processes. In recent years, the amount of lactic acid obtained by biotechnological methods has increased. The highest cost in the traditional process of lactic acid production by carbohydrate fermentation lies in the separation steps that are needed to recover and purify the product from the fermentation broth. To reduce these costs, many studies on lactic acid separation have been conducted using different separation techniques. In this chapter the purification of lactic acid by means of membrane techniques and ion exchange are considered. The separation efficiency of nanofiltration membranes in the treatment of ionic solutions is explained by the combination of size and charge effects. A strong influence of pH on lactic acid transport through nanofiltration membranes is observed. The pH value affected both rejection and permeate flux. Rejection increased with pH while flux decreased with this variable. At the acidic pH required to convert lactate into undissociated lactic acid (pH 3.0), the recovery of lactic acid is low (around 50 %) and similar to that of the inorganic salts present in the fermentation broth. Therefore it should be improved in order to reduce lactic acid losses and increase purity. Special interest has received the recovery of lactic acid using bipolar membrane electrodialysis due to the possibility of obtaining lactic acid and caustic soda simultaneously from pretreated fermentation broths. In the second part of the chapter a review on different processes for lactic acid recovery based on ion exchange is performed. Most of the works are focused on the recovery of lactate ions from the broth by means of anion exchangers However, the separation process proposed in this chapter is an ion-exchange process that makes it possible to remove undesired ions from the fermentation broth, while lactic acid remains in solution. The final purity of lactic acid was higher than 99%. Losses were estimated to be lower than 15 %.

Keywords: Lactic acid, purification, membrane technology, ion exchange

INTRODUCTION

Lactic acid is one of the most widely used food chemicals. It is mainly utilized as a preservative and acidulant in the food industry and in the last few years a great deal of interest has been devoted to employing it as a source for the production of biodegradable polylactic polymers, such as polylactic acid (PLA) [1]. The manufacture of PLA will probably be responsible for a significant increase in lactic acid production in the next years. Nowadays, the market for lactic acid is around 300,000 tons per year. Production is expected to increase in the coming years, but the rate of growth depends on the production costs [2].

Lactic acid can be manufactured either by chemical synthesis or by carbohydrate fermentation, but around 90% of world production follows the second way. From the point of view of sustainable development and human health, the fermentative process is preferred. The chemical synthesis route produces only a racemic mixture of lactic acid while the fermentation process can obtain an specific stereoisomer of lactic acid (L(+)-lactic stereoisomer is preferred because it is metabolized by the human body). Overtaking D(-) or DL-lactic acid causes metabolism disorders [3].

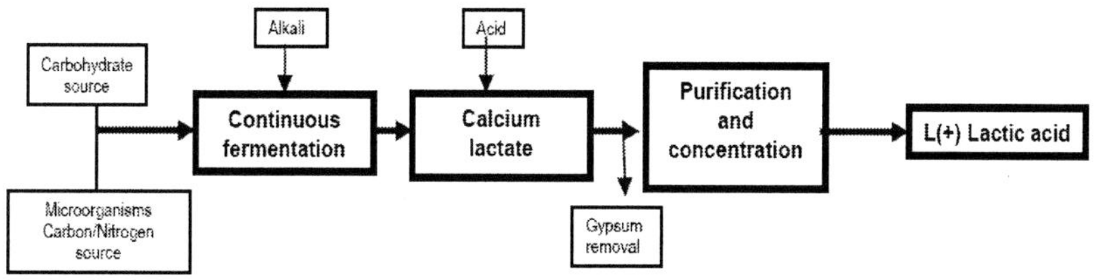

Figure 1. Simplified scheme of the conventional process for lactic acid production.

The composition of the fermentation broths depends on the carbohydrate source, concentration and type of microorganism used as well as on the operation mode chosen to perform the fermentation process (continuous/discontinuous). Lactic acid concentration in the broths can vary between 40 and 100 $g.L^{-1}$, with productivities between 1 and 14 $g.L^{-1}.h^{-1}$ [2].

In the conventional lactic acid fermentation process (see figure 1), which is typically performed at pH values between 5 and 6, an alkali (usually calcium carbonate) is used to neutralize the pH in order to avoid very acidic conditions (pH lower than 4), which negatively affect the fermentation. Moreover, lactic acid fermentation is end-product inhibited. Other chemicals, such as ammonium and sodium hydroxide can also be used, but they are more expensive. Once the fermentation is finished it is necessary to separate the organic acid salt (calcium lactate) and to convert it into lactic acid by reducing the pH (with sulphuric acid usually). Thus gypsum is obtained as by-product (approximate 1 ton of gypsum per ton of lactic acid produced) [4,5].

To recover lactic acid in the conventional process a great number of separation steps (precipitation, distillation, filtration, acidification, crystallization, etc.) are necessary [6,7]. These operations account for around 50% of the total production costs, introduce further complexities in the process, show low yield and productivity, produce a large amount of co-products and a negative impact on the environment and generate solid wastes; thus, alternative processes are necessary [7-9].

As an alternative to the traditional purification processes, different new technologies have been investigated in order to obtain purified lactic acid at low cost (reactive extraction, membrane technology, ion exchange, electrodialysis, etc.) [6,8-13]. Most of them have been compared by different authors [14,15]. Environmentally friendly and cost effective techniques are required for the production and downstream processing of lactic acid at as low cost as possible.

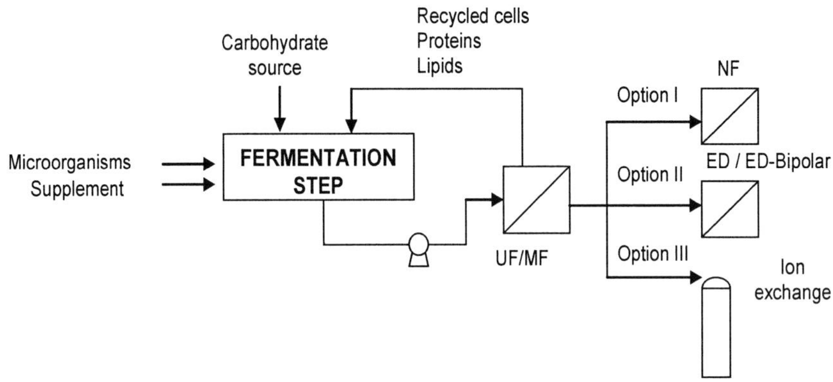

Figure 2. Different approaches for downstream processing of lactic acid.

This chapter reviews some new approaches to perform lactic acid downstream processing by means of membrane technologies, including pressure driven and electrically driven processes, and ion exchange. The possible schemes considered in this chapter to perform lactic acid recovery from fermentation broths are shown in figure 2.

Membrane Technologies Applied to Lactic Acid Purification

Membrane technologies (MT) are well known since a few decades ago. The development in porous organic and inorganic materials, the new methodologies to control pore size and porosities, the reduction in prizes and the development of a number of industrial applications are the reasons for their growth.

The easiest way to classify the main membrane technologies is according to the driving force used in the process. Thus, they can be classified into pressure driven membrane processes (Microfiltration-MF, Ultrafiltration-UF, Nanofiltration-NF and Reverse Osmosis-RO) and electrically driven membrane processes (Electrodialysis-ED).

All of them have been studied in the downstream processing of lactic acid. There are a number of excellent books and papers that review the fundamentals and applications of membrane processes, membrane manufacturing technologies and mathematical models to predict membrane behaviour and performance [16-21].

Pressure Driven Membrane Processes Applied to Lactic Acid Recovery

With the objective of simplifying the complex conventional lactic acid recovery process, many researchers have applied MT directly to the fermentation broths. All theses studies can be classified in two groups: those where MT (especially MF and UF) is used to recycle cells and macromolecules (non fermented carbohydrates, proteins, etc.) and those that intend to recover lactic acid directly from the fermentation broths (NF, ED). The success of these two approaches has been different.

Fermentation reactors coupled with membranes (MF and UF) represent a well-established technology (MBR) at industrial scale, which has been often used to treat industrial and urban wastewaters as well as in several applications in the food industry [21-34]. In lactic acid production, the advantages of MBR systems are the possibility of working in a continuous mode (figure 2) with continuous cell recycling and removal from the fermentation reactor of lactic acid and other components (salts, non-protein nitrogen, etc.) in the permeate reducing the acidity that negatively affects the microorganisms when cell concentration increases above 10 $g.L^{-1}$ [4,5]. Membrane behaviour, especially permeate flux, selectivity and fouling, strongly depends on broth pH, type of membrane (MF with pore sizes between 0.1 – 0.2 μm and UF with molecular weight cut-offs (MWCO) between 50 and 300 kDa have been reported) and type of membrane modules. In MBR applied to lactic acid production, important problems of membrane fouling, low permeate flux and short periods of operation (high cleaning frequency) have been described in most of the works [35]. However, when the MBR system is properly operated and the feed stream is less complex in its composition, the efficiency of MT notably increases [13,14]. When NF is going to be applied to lactic acid purification some points must be taken into account: feed composition (presence of ionic and nonionic organic and inorganic compounds), feed pH (that affects the ratio of dissociated to undissociated lactic acid) and membrane characteristics (mainly the isoelectric point-IEP). Nanofiltration is a pressure driven membrane process in which membrane pore radius varies between 1 and several nanometers and that is usually used to separate molecules smaller than 1000 Da (150-250 Da preferably). This technique is being used in a wide range of applications in food industry and wastewater treatment as well as in downstream processes where organic acids are present in the form of salts and non ionic organic compounds, many of them coming from fermentation processes [37-40].

Table 1. Membrane technologies (NF-RO) applied to lactose and lactic acid recovery

Feed	Process	Membrane characteristics	Operating conditions	Average flux	Lactic acid or Lactate rejection (%)	Ref.
Model solutions (electrolyte + (lactate /glucose)	NF	Desal® kDa (Osmonics) Flat sheet. 0.0135 m^2	0.2-1.8 MPa pH = 6-7 25°C, 1.33 $m.s^{-1}$	Maximum 90 $L.h^{-1}.m^{-2}$	80-95% 30% reduction when increasing NaCl and Na_2SO_4 concentration	[59]
Ultrafiltered whey	NF	Desal® 5kDa Spiral wound 2.51 m^2	2.5 MPa pH: 6.5, 22°C 50 $g.L^{-1}$ lactose	Maximum 30 $L.h^{-1}.m^{-2}$	80-90% Membrane negatively charged	[36]
Fermentation broth clarified by UF (15 kDa)	NF	AFC80® Polyamide Tubular organic membrane R_{NaCl} = 80%, 0.72 m^2	1 - 4 MPa pH: 2.7 - 6.0 20 – 40°C	Maximum 80 $L.h^{-1}.m^2$	Between: 50% (pH 2.7) 90% (pH 6)	[13]
Fermentation broth	NF + RO	Several NF membranes (GE Osmonics) MWCO = 100 - 400 Da	1.4 – 2.8 MPa pH = 5.5 37°C	30 - 40 $L.h^{-1}.m^{-2}$	Maximum lactose rejection 97% Minimum lactose rejection 15%	[57]
Fermentation broth clarified by UF (15 kDa)	NF	DK2540C® (Filtr. Eng.) Polyamide. Spiral wound $R_{lactose}$ = 96-100%; 1.8 m^2	1 - 4 MPa pH: 2.7 - 6.0 20 – 40°C	Maximum (4 MPa) 220 $L.h^{-1}.m^{-2}$	Between:10% (pH 2.7) – 90% (pH 6)	[13]
Fermen tation broth (from starch residues)	NF	DK (Osmonics). Polyamide Flat-Sheet, 0.0140m^2 IEP = 4 MWCO =150 – 300 Da	0.2 – 2 MPa 25°C 1.3 $m.s^{-1}$	35 $L.h^{-1}.m^{-2}$ (2.0 MPa)	Around 20%	[55]
Lactic acid aqueous solution Conc.:6.8–340 gL^{-1}	NF	Inorganic (Zirconium) dynamically formed Pore radius: 2.6 -3.2 nm	1.5 – 5.5 MPa pH: 4 – 8 1 – 2.6 $m.s^{-1}$	Not available	Very low lactic acid rejection at high concentration and low transmembrane pressures	[12]

Table 1. (Continued)

Feed	Process	Membrane characteristics	Operating conditions	Average flux	Lactic acid or Lactate rejection (%)	Ref.
Model solutions (sugars, lactate, salts)	NF	Desal 5® (Osmonics) Polyamide/Polysulphone IEP = 4, 0.0135 m^2 R_{MgSO4} = 98%	0.2 – 2 MPa 25°C 0.33 – 1.33 $m.s^{-1}$	6 – 10 $L.h^{-1}.m^{-2}.bar^{-1}$	Lactate retention of about 80% Strong influence of glucose and salts present in the feed	[55]
Fermentation broth (sweet whey)	NF	R76A® (Millipore) Spiral wound 0.30 m^2 MWCO: 300 Da	1 - 3 MPa pH: 3.3 – 6 30°C 0.75 $m.s^{-1}$,	20.2-3.9 $L.h^{-1}.m^2$	$R_{Lactose}$: 97% (constant with pH) $R_{Lactate}$: 12-35% (pH:3.3) $R_{Lactate}$:36-72% (pH: 6)	[35]
Pretreated Fermentation broth	RO	NTR7199 (Nitto Denko) Flat sheet Polyamide composite	1– 7 MPa pH: 4.2-5.1 Feed conc.:3%	28 – 20 $L.h^{-1}.m^{-2}$ (pH 4.2-10.1) 75 $L.h^{-1}.m^{-2}$(pH .2.2)	$R_{Lactate}$>83%	[54]
Model lactic acid solution (1% w/v)	RO	HR95® (DDS) Thin-film composite 0.036 m^2 ------------------------------ CA995® (DDS)	1- 4 MPa pH: 2.8 - 4.9 4 -10 $L.min^{-1}$	Not available	$R_{Lactic\ acid}$: 99.5% (pH:2.88) $R_{Lactic\ acid}$: 97.4% (pH:4.93) ---------------------------- $R_{Lactic\ acid}$: 84.2% (pH:2.88) $R_{Lactic\ acid}$: 73.0% (pH:4.93)	[52]
Model lactic acid solution (1% w/v)	NF	DDS HC50 (NIRO) Thin-film composite ------------------------------- DDS CA960 (NIRO) Celulose acetate	1- 4 MPa pH: 2.8 - 4.9 4 -10 $L.min^{-1}$	Not available	$R_{Lactic\ acid}$: 52.6% (pH:3.45) $R_{Lactic\ acid}$: 49.7% (pH:4.93) ------------------------------------- $R_{Lactic\ acid}$: 33.2% (pH:2.88) $R_{Lactic\ acid}$: 67.5% (pH: 4.93)	[52]
Ultrafiltered Fermentation broth	NF	NF40® (FILMTEC) Spiral wound 6 m^2	3-5 MPa 37°C 4.8 $m^3.h^{-1}$	11-60 $L.h^{-1}.m^{-2}$	Lactic acid rejection about 40%	[54]

In NF, prediction of membrane permeability and selectivity is more complex that in MF and UF due to the significant diffusion and electrostatic effects that can play an important role depending on the pH, ionic strength, membrane characteristics, solute charges, etc. Different models to estimate permeate flux and membrane rejections can be found in the literature. Most of them have been applied to diluted model solutions of salt mixtures, salts/ionic organic compounds mixtures or organic non ionic/organic ionic acids/salts mixtures, but the models are less efficient when applied to concentrated streams due to the influence of concentration polarization and membrane fouling [41-51]. Direct nanofiltration of fermentation broths theoretically allows to obtain lactic acid (molecular weight of 90.08 $g.mol^{-1}$) and other salts in the permeate stream as well as protein nitrogen from small peptides and aminoacids, being retained cells, proteins and the rest of broth components. Nanofiltration (and RO) coupled to acid lactic fermentation has been investigated by different researchers [34] and some information can be found in Table 1. However, this technique presents great difficulties due to the broth complexity, high cell concentration and strong membrane fouling [35,52-57]. When fermentation broths are fed to a nanofiltration membrane in a continuous way, non fermented carbohydrates must be separated from lactic acid/lactate. Some authors [55] have demonstrated that carbohydrates (such as glucose or lactose) are more easily transmitted through the membrane in the presence of lactic acid, thus membrane selectivity is too low. It has been established that glucose, for example, is less hydrated when NaCl is added to the solution [58], showing lower retention than expected for NF operations. This behaviour of higher transmission of non ionic solutes in the presence of a charged one has been also found for different salts (different concentration and type of salts – monovalent/divalent), being the non ionic organic compounds even less rejected when multivalent electrolytes are present [42,43,59,60].

Similar results have been obtained by other authors [61] with different membrane materials and solutes. Organic molecules do not seem to be retained only by sieving mechanisms and some researchers have found certain influence of polarity and charge effects [62]. Organic molecules with a high dipole moment have lower retention compared to non-polar molecules. This behaviour is not dependant on the charge sign (positive or negative) of the membrane and can be explained by the electrostatic interaction between the dipole and the membrane surface. When the NF membrane has a small pore size, its electric charge has low effect on molecule retention, but this influence increases if the pore size is larger. The presence of salts can increase the membrane charge density due to a higher concentration of counter-ions in the electric double-layer at the pore surface [42,46]. Also, changes in the effective pore size can occur due to swelling of the membrane material if there are strong repulsion forces between pore walls. This theory has not been completely demonstrated since the same behaviour has been found in the case of ceramic membranes (which are not expected to swell) [63].

Different NF membranes have been tested to separate lactic acid from unfermented lactose (rejection 97±1%) in cheese whey fermentation broths at pH 5.5 [64]. The obtained permeate, composed mainly by lactic acid, salts and water, was then concentrated by means of a RO membrane. The effect of transmembrane pressure (between 1.4 and 2.8 MPa) and lactose/lactic acid ratio was investigated. Lactose retention was observed to be higher than that of lactic acid for all the experimental conditions and membranes tested (MWCO between 175 and 1000 Da), obtaining higher differences of retention (96,8% for lactose and 43.7% for lactic acid) for thin composite membranes with 98% $MgSO_4$ retention. However, lactic acid recovery was too low.

Another important aspect to take into consideration is the membrane isoelectric point (IEP). That value is related to the material the membrane is manufactured from and to the concentration and type of electrolyte present in the feed [65,66]. The most often used NF membranes show IEP between 3 and 6, thus, nanofiltration membranes are usually negatively charged in neutral or alkaline conditions and positively charged in acidic conditions [45]. When a NF membrane is negatively charged, solutes with high anionic charge densities are more effectively rejected. For this reason, divalent ions usually show higher rejection in NF membranes than monovalent ones; additionally non-ionic solutes are less rejected if their size is smaller than membrane pore size. As a result, the charged NF membrane can reject ions much smaller than the membrane pore size [50,67,68].

Another important point to take into account when organic electrolytes are involved in separation or fractionation processes, is the dissociation equilibrium. In lactic acid fermentation pH must be maintained between 5 and 6 in order to avoid inhibitory effects. The degree of acid lactic dissociation is dependant on the pH value and can be estimated by means of the Henderson-Hasselbalch equation:

$$CH_3CHOHCOOH \leftrightarrow CH_3CHOHCOO^- + H^+$$

$$pH = pK_a + Log\frac{\left[CH_3CHOHCOO^-\right]}{\left[CH_3CHOHCOOH\right]}$$

The pK_a for lactic acid at 25°C is 3.86. At pH 2.7 only 6.5% of the lactic acid is dissociated, whereas 93.3% of the lactic acid is dissociated at pH 5 and 99.3 at pH 6. Additionally, the pK_a of lactic acid decreases with temperature (at 40°C it is estimated to be 3.67). The degree of lactic acid dissociation is important because NF rejection depends on it. At the pH of the fermentation broth, most of the lactic acid is dissociated, what improves the fermentation productivity because the undissociated acid has a strong inhibitory effect [34]. Figure 3 shows the percentage of dissociation at different pH values and at 25 and 40°C.

When pH raises lactate/lactic acid ratio increases according to the Henderson-Hasselbalch equation, thus the rejection of the NF membranes increases as well, no matter the membrane MWCO and charge [13]. At highe pH values the membrane is negatively charged and electrostatic repulsion between lactate and the membrane surface is significant due to the Donnan exclusion effect. The greater the membrane charge density is the higher the lactate rejection. Similar conclusions have been obtained by other authors [35,52] working with different membrane configurations and similar MWCO. Additionally, as a consequence of lactic acid dissociation, the pH value affects the osmotic pressure, which increases at higher pH values, causing lower permeate fluxes [13].

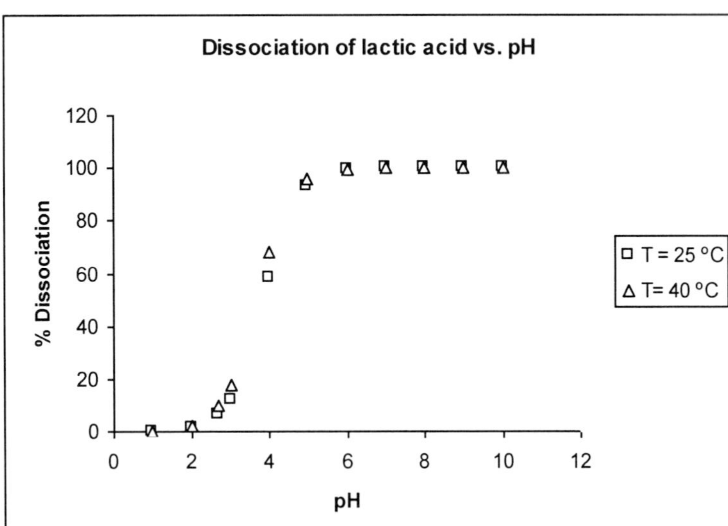

Figure 3. Percentage of dissociation of lactic acid vs. pH at 25 and 40°C.

The effect of pH is in fact one of the most studied aspects in lactic acid purification by NF [13,29,52]. The electrostatic effect is a limiting factor in the recovery of lactic acid by NF. Depending on the membrane charge, rejection can be higher or lower than expected on the basis of molecular weight or another size parameter.

Some authors [69] reported that charge can also change the effective membrane pore size, affecting membrane flux and rejection. At high pH values the effective pore size decreases because the charged carboxylic groups adopt an extended chain conformation due to electrostatic repulsion, reducing the pore size (or pore volume) and thereby causing decreased flux and increased salt rejection.

Lactic acid concentration has also a great influence on rejection and permeability, although the general trend described before is maintained. When lactic acid concentration increases the Donnan exclusion effect was attenuated due to the more significant concentration polarization effect.

Most of the models developed to describe NF processes that are based on Donnan exclusion effect and other electrostatic interactions do not fit the experimental data well when concentrated solutions are treated (solute concentration greater than 20 or 30 $g.L^{-1}$).

Due to the great influence of solutes-membrane interactions previously described, the selection of the NF membrane is important to reach good results. The most important properties are: molecular weight cut-off (MWCO), porosity, surface roughness, hydrophobicity, surface charge and water permeability [40]. In the case of filtration of organic compounds (no charged compounds), hidrophobicity of the top layer of the membrane has a very negative influence on flux decline (fouling) [70].

Regarding the NF process conditions, all the research works showed that membrane rejection increases with transmembrane pressure due to the increase in solute concentration at both sides of the membrane [35].

Increasing the cross-flow velocity causes an increase in permeate flux (due to reduction of concentration polarization effect). Generally, permeate flux and rejection show reverse trends. This behaviour has been described by several authors in different conditions (modified inorganic membranes, artificial lactic acid solutions) [12].

As general conclusion, low fluxes and selectivities were obtained when fermentation broths were directly treated by NF. Therefore, further development is necessary so that the process can be economically accepted at industrial scale.

Electrodialysis Based Processes Applied to Lactic Acid Purification

Electrodialysis is a very promising method to separate ionic compounds.. It has received a lot of attention in the last years thanks to the rapid improvement in membrane materials. Electrodialysis is employed to remove salts from solutions or to concentrate ionic substances, but in the last years ED and ED based technologies have been also used to obtain and to separate organic acids due to the possibility of simultaneous supply of H^+ and OH^- ions without the introduction of chemicals [71-73]. Nowadays, a number of different ED based processes have been developed: Conventional ED, ED with bipolar membranes (EBM), electrometathesis (EMT), electrodeionization (EDI), electro-electrodialysis (EED), eletro-ion injection-extraction (EIIE), etc.; and some of them have been tested for lactic acid/lactate recovery. Several articles have described the fundaments and applications of these ED based processes to organic acid production [74-77].

One basic concept in ED is the limiting current density (LCD), which is the maximum current that may pass through a given cell pair area without detrimental effects. If the limiting current density is exceeded, the electric resistance of the diluate (feed) will increase and water dissociation may occur at the membrane surface, causing efficiency losses and changes in the solutions pH. LCD can be affected by a number of parameters (electric power requirements, lifetime of membranes, product quality, etc.) and its value can change with the working conditions. The best method to perform the empirical estimation of LCD is the current-voltage method (Cowan and Brown method) [78]. Experimental LCD corresponds to a change in the slope of the curve obtained when cell resistance (Ohm) is plotted vs. reciprocal current (A^{-1}).

There are two main ED technologies applied at industrial scale that can be used for lactic acid downstream processing. The first technology is the conventional ED processes, which consists in a series of anion and cation exchange membranes arranged in an alternative pattern between an anode and a cathode (see figure 4). In this configuration the lactate salt solution (sodium lactate in figure 4) mixed with other non ionic compounds is pumped through the stack in which an electric potential is established.

Cations migrate towards the electrode of the opposite sign (cathode) and they pass trough a cationic exchange membrane (negatively charged) to the next compartment in which a positively charged membrane (anionic exchange membrane) rejects them due to the Donnan exclusion effect.

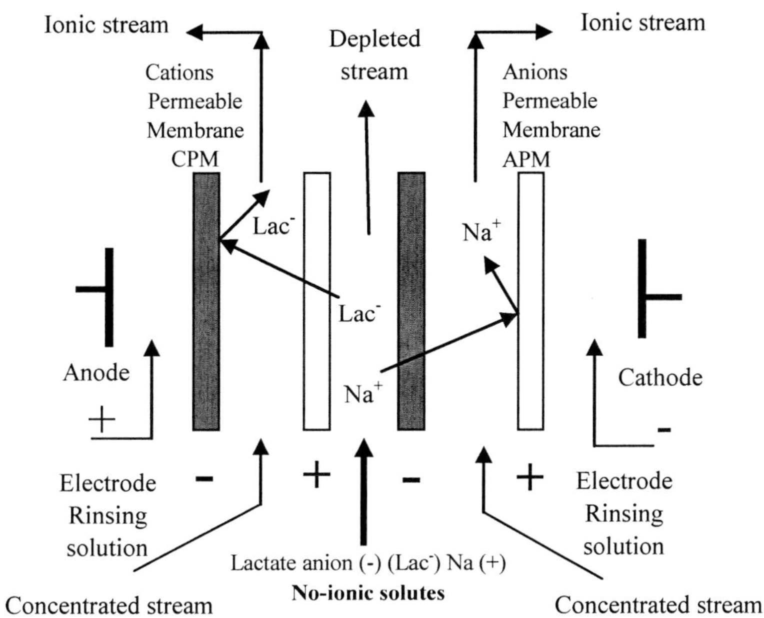

Figure 4. Conventional ED process applied to lactate purification.

In a similar way, lactate anions pass through a positively charged membrane and they are rejected by a negatively charged membrane. The overall result is an increase in the ion concentration in alternate compartments (concentrate stream or brine) while the other compartments simultaneously become depleted (diluate stream). Moreover, selective anion or cation exchange membranes have significantly different transport properties for monovalent and multivalent ions, so that the ions can be separated on the basis of their charge, the same as happens in a NF process [48].

Conventional ED can be used to separate lactate salts from non ionic compounds present in fermentation broths or other sources [79,80]. Current efficiency is slightly higher than 70% and lactic acid recovery higher than 90%. Conventional ED can be combined with the so-called bipolar membranes constituting a bipolar membrane electrodialysis (EBM) stack. Bipolar electrodialysis involves water splitting within the bipolar membrane, which is composed of a cation-exchange membrane laminated together with an anion-exchange membrane. Therefore protons and hydroxyl ions are efficiently generated and acids and bases are produced from salt solutions with high energy efficiency.

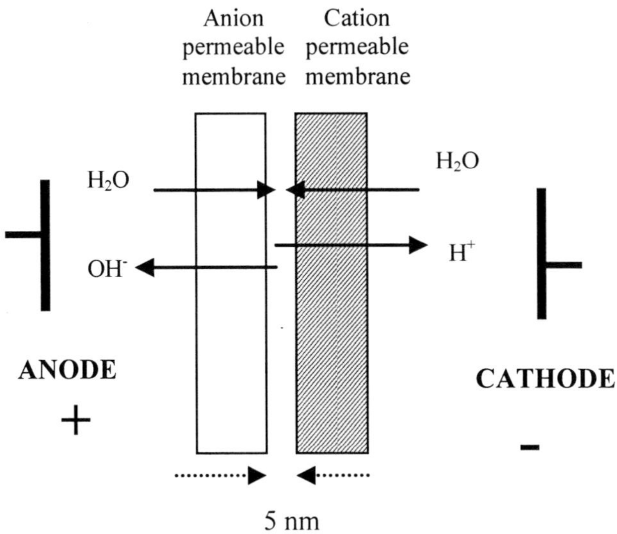

Figure 5. Structure and operation of bipolar membranes.

Bipolar electrodialysis provides a potentially attractive complement to lactic acid fermentation, enabling the separation, purification and concentration of salts and converting these salts into acids and bases without producing effluents containing high concentrations of salts or gypsum, thereby avoiding their discharge to the environment. Additionally, the alkali produced can be recycled back to the fermentation process [81].

In ED with bipolar membranes (figure 5), when an electric potential is established in an aqueous media water is dissociated generating hydroxyl ions (OH^-) and protons (H^+) that move to the opposite electrodes. The H^+ and OH^- ions can react with anions (organic or inorganic) or cations to obtain acids and alkalis simultaneously.

EBM stacks can be designed in different ways (two or three compartment configurations) depending on the final result desired. In case of two compartments, the stack configuration depends on if the objective is to obtain pure acid (organic or inorganic) or to produce a concentrate alkaline stream. Figure 6 shows the most appropriate stack configuration when the obtention of lactic acid is the objective and a mixture of lactic acid/lactate and non ionic compounds is feed to the system. In a three compartments EBM (see figure 7) monopolar membranes (cationic and anionic) are installed in combination to bipolar membranes forming an ED stack that can be used to produce acids and alkalis from the corresponding salts.

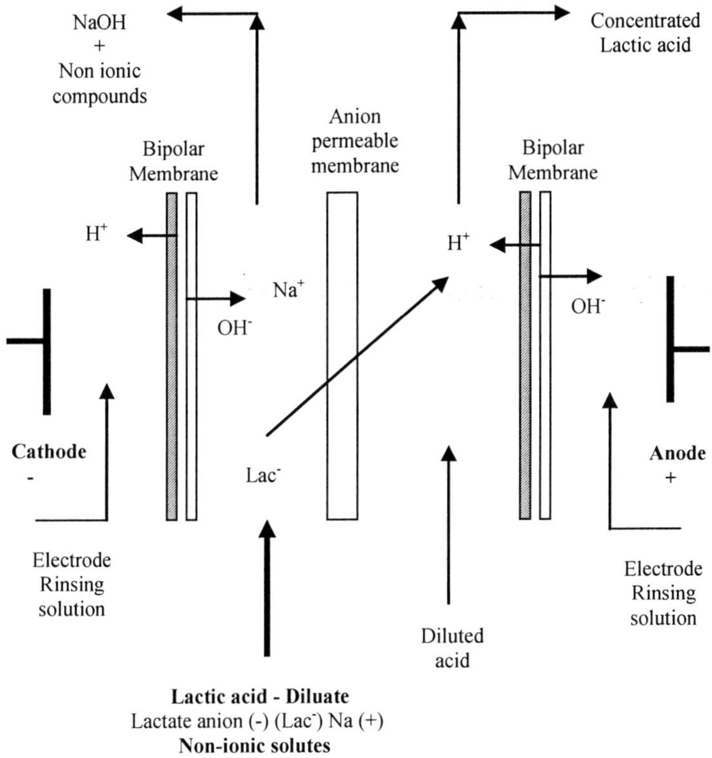

Figure 6. Two compartments EBM (lactic acid recovery).

The ions $H^+$and OH^-generated by the bipolar membrane are combined with lactate anions and Na^+, respectively, obtaining lactic acid and NaOH at the same time. ED based processes have been considered as efficient and environment friendly methods for lactic production and purification, however the commercialization of bipolar membranes at industrial scale has been very limited until now due, mainly, to the permanent damage of the membranes specially when high current densities are applied.

All ED based processes used to recover and purify lactic acid from fermentation broths need intensive pre-treatments to ensure stable operation and process performance. Microfiltration or ultrafiltration are commonly used to remove cells, particles, colloids or other suspended solids and keep the membranes free from undesirable fouling [82,83]. Some authors included another step of decolouring using activated carbon [11] in order to remove dyes or pigments that can colour the final product or interact with membrane surface.

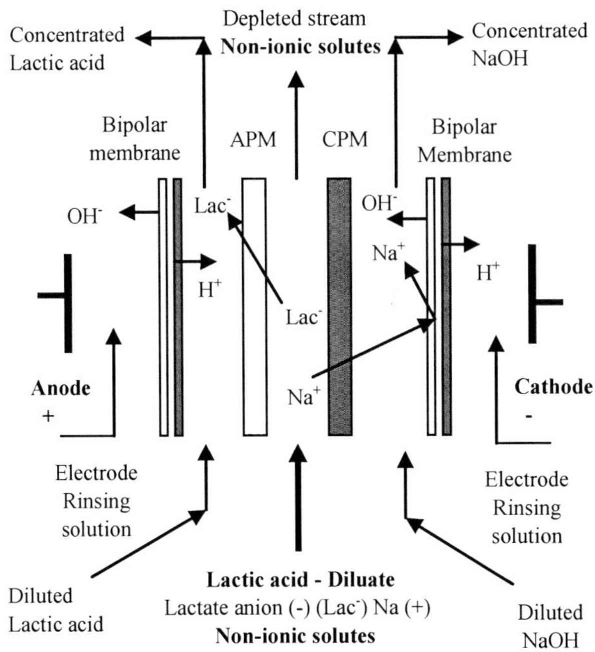

Figure 7. Three compartments EBM (lactic acid recovery).

Finally, the removal of multivalent inorganic ions is recommended to prevent scale formation, especially when bipolar membranes are used [84]. Chelating resins are used in some cases to reach this objective [85, 86].

ED coupled to fermentation has been investigated by several authors [15,87]. In the case of lactic acid production, ED technologies are interesting because lactic acid can be continuously removed from the fermentation reactor, thus maintaining the pH of the broth. ED based process does not require the addition of acid or alkali to convert lactate into lactic acid and this is one of the most interesting advantages of this technology. Some authors [88] have obtained good results (95% lactic acid recovery) in the treatment of previously microfiltered fermentation broths (synthetic), which contained high levels of foulants (residual glucose, proteins, etc.). They used a three compartment configuration and one stage ED. Li et al., 2004, have tested an special arrangement (electrokinetic bioreactor) of anion and cation exchange membranes combined with a bipolar membrane to separate lactate from a real fermentation broth (which contained glucose). The stack is able to remove lactate from the fermentation broth, while the pH is itself controlled by the combination of the H^+ and OH^- ions generated by the bipolar membrane. However, direct extraction of lactic acid from fermentation broths poses some problems, such as the possibility of cell damage in contact with charged membranes [89]. Another problem also described is the leakage of carbohydrate (glucose) into the system [90]. Even though the carbohydrate is a non charged solute, it can go to the recovery solution with electro-osmosis of water. This glucose leakage reduces the fermentation efficiency.

Most of the works that apply ED technologies to lactic production follow a two-steps scheme. First, lactic acid from the fermentation reactor is concentrated and purified by conventional ED or NF removing inorganic salts. In the second step, bipolar membranes (EBM) are used to obtain lactic acid and an alkali (usually NaOH) that can be recycled to the fermentation tank. By using this scheme some authors [86] have obtained up to 151 $g.L^{-1}$ lactic acid at the end of the second step (EBM) with a total energy consumption of 1.5kW.h per kg of lactic obtained. However, special attention must be paid to the first step in which lactic acid losses can be significant if the ED process is not performed at optimum pH.

Some authors have demonstrated that lactic acid losses are higher when this organic acid is dissociated to a great extent, passing through the ED membranes the same as the inorganic salts [91,92].

Madzingaidzo et al., 2002 [11] provided detailed data on the treatment of a cell free sodium lactate solution processed in two steps as well. The conventional ED process showed a good performance with lactate flux of up to 300 $g.m^{-2}.h^{-1}$, a current efficiency of 80% and a final concentration of lactate (which ranged from 120 to 190 $g.L^{-1}$) higher than those previously reported. Energy consumption was clearly below 1 kW h per kg lactate. In the second EBM step, the final concentration of lactic acid obtained was around 160 $g.L^{-1}$ and the maximum NaOH concentration was of 100 $g.L^{-1}$. Energy consumption in this second step ranged between 0.6 and 1 kW.h per kg lactate.

Most of the works published have only considered model solutions, low salt contents or simple mixtures. Electrodialysis of real fermentation broths or streams containing salts or other compounds at high concentrations are not so common. Thang et al., 2005 [80] investigated the recovery of lactic acid from a complex silage juice by using the two steps scheme. Grass silage was previously ultrafiltered and then, the permeate was submitted to a two steps ED at two different pH values (6.7 and 2). Current efficiency of both steps was 67% and 88%, respectively and the amount of lactic acid recovered was similar for both steps (around 95%). The energy consumption was 2.11kW.h.L^{-1} for the first step and 1.08 kW.h.L^{-1} for the second one.

Table 2. Comparison of different alternatives to produce lactic acid from fermentation broths by means of ED

Process	Lactic acid recovery (%)	Energy consumption ($kW.h.kg^{-1}$)	Carbohydrate Rejection (%)	Divalent salts rejection (%)
NF+THCEBN	96	1	61	82
Two stages ED	98.9	0.9	84	60
TWCEBN	74	3.4	100	100
THCEBN	99	1.4	98	80

NF: Nanofiltration.
THCEBN: Three compartments EBM.
TWCEBN: Two compartments EBN.

Finally, some authors have compared different alternatives to produce lactic acid from fermentation broths (at laboratory scale) [84, 88, 93]. The main results are shown in Table 2. As can be observed, the best results were obtained by using three compartment configurations and electrodialysis with bipolar membranes.

The combination of different technologies is possible to reach the objective of isolation, purification and concentration acid lactic from a fermentation broth. Each technique has advantages and disadvantages and probably the best solution from the point of view of degree of lactic acid recovery, purity and efficiency could be an integrated process that combines the best of each technique.

RECOVERY IF LACTIC ACID BY ION EXCHANGE

Ion exchange is a separation process widely used in bioseparations, which has been considered by many authors to perform the recovery of lactic acid from fermentation broths. It consists of the reversible interchange of ions between a solid (ion exchange material) and a liquid [94]. The exchange reaction can be represented as follows:

$$AR + B^{n\pm} \leftrightarrow BR + A^{n\pm}$$

where R is the ion exchange material (usually an ion exchange resin), A is the ion that the resin contains and B is the ion present in the solution.

There are two groups of ion exchangers:

Cation Exchangers. They have acidic groups that can exchange organic or inorganic cations with the solution or replace them with protons. Strong acid resins show a chemical behavior similar to that of a strong acid. The ionizable group is a sulfonic acid group. These resins are highly ionized in both the acid ($R\text{-}SO_3H$) and salt ($R\text{-}SO_3Na$) form. In a weak acid resin, the ionizable group is a carboxylic acid. These resins behave similarly to weak organic acids that are weakly dissociated.

Anion Exchangers. They have basic groups that can exchange organic or inorganic anions with the solution or replace them with the hydroxyl ion. Strong base resins are highly ionized and can be used over the entire pH range. However, the degree of ionization of weak base resins, like weak acid resins, is strongly influenced by pH.

In most of the cases, both at laboratory and industrial scale, the ion exchange operation is performed in a column. Thus, the solution is passed through a column containing a bed of ion exchange resin. At the beginning, the exchange occurs in a short area at the inlet of the column. As the process continues, the area of ion exchange moves upwards until it reaches the top of the column.

This corresponds to the *breakthrough point.* The new ions attached to the column can be recovered by elution. After the feed solution is processed to the extent that the resin becomes exhausted, the resin must be regenerated back to its original form using the appropriate chemicals.

Different approaches have been proposed in the literature to recover lactic acid from fermentation broths by ion exchange. Most of the papers are focused on the recovery of lactate ions from the broth by means of anion exchangers. Lactate or lactic acid are then eluted by means of the appropriate substances [95,96].

A novel approach consists of the utilization of a cation exchanger followed by an anion exchanger to remove both undesirable cations and anions present in the broth, while lactic acid remains in solution [9,97]. Different types of feed streams and different types of ion exchange resins have been considered in the literature.

As lactic acid fermentation is end-product inhibited, the coupling of fermentation and ion exchange to continuously remove the lactic acid from the bioreactor can increase the overall production yield. Moreover, the pH can be maintained at the optimum values, which typically range between 5 and 6.

Recovery of Lactate Ions by Anion Exchangers

Many processes have been proposed and many ion exchange resins have been tested by different authors [92,96]. Some of the authors used synthetic lactic acid solutions as feed streams, others used artificial fermentation broths, while others used real fermentation broths from different sources (whey, cereals, molasses, wood, paper sludge, cassava bagasse, apple pomace, etc.).

In some of the works the capacity of the resins at different experimental conditions as well as the equilibrium isotherms and the ion exchange kinetics were investigated in batch conditions by mixing a certain amount of dry resin with aqueous lactic acid solutions [98-113]. High ion exchange capacity was observed for most of the resins tested. Moreover, capacity was observed to decrease with temperature [98,103,112]. Some authors also investigated the effect of pH on the capacity of the resins. In most of the works, the weak base resins were used in the free base form and the strong base resins in the OH^- form. When weak or moderate base ion exchangers are used, pH plays an important role. When pH values are greater than the resin pK_a, the ion exchanger is unionized and it does not accept anions [114]. Therefore, the capacity of these anion exchangers was observed to decrease with pH [100]. However, strong base ion exchange resins are much less affected by pH. According to Evangelista et al., 1994, weak base resins in the free base form can be used up to pH 3.8, moderate base resins up to pH 4.5 and strong base resins are not affect by pH up to pH 5.8 (the highest pH considered by the authors) [100]. Weak and moderate base exchangers in the free base form prefer undissociated lactic acid. At low pH values most of the authors found that weak anion exchangers are more suitable for lactic acid recovery than strong ones as they show higher capacities [106,107]. Moldes et al., 2003 tested several weak and strong anion exchange resins. The resins were tested in the free base and OH^- form, but they were also tested in the Cl^- from. They reported that, at pH 4.85, which is higher than lactic acid pK_a, weak base resins in the Cl^- form presented the highest capacities. However, those resins in the free base form showed much lower capacities. On the other hand, strong base resins in the OH^- form showed higher capacities than in the Cl^- form and than those of the weak base resins in the free base form. These authors indicated that strong base resins usually have low affinities for the OH^- ion, therefore they exchange it easily. Moreover, strong base resins in the OH^- form can interact with both dissociated and undissociated form of lactic acid, but weak base resins in the free base form can only interact with the undissociated form [106]. The same conclusions were obtained by John et al, 2008 [110].

Therefore it seems that weak base reins in the Cl^- form are the most appropriate anion exchangers for lactic acid recovery. Moldes et al., 2003 also observed that the capacities of the resins with macroreticular structure are slightly higher than those of the gel-matrix resins [106].

Regarding the equilibrium isotherms, most of the authors found that the ion exchange of lactic acid can be described by a Langmuir type equation [98,100,102,103,105,107,112]. However, Cao et al., 2002 obtained different types of isotherms at pH 2.0 and 5.0, respectively, when lactic acid was recovered by the anion exchange resin Amberlite IRA-400. At pH 5.0 the isotherm obtained was a Langmuir form, while at pH 2.0 it corresponded to type II adsorption [105]. Lihua et al., 1995 and Lin et al., 1998 reported that simultaneous ion exchange and physical adsorption occurred in the recovery of lactic acid with the weak basic D-354 resin [98,103]. However, Moldes et al., 2003 indicated that the predominant mechanism is ion exchange and not adsorption in the recovery of lactic acid by means of several strong and weak anion exchange resins [106]. Some authors also studied ion exchange kinetics [102,106,112]. Most of them observed first order kinetics, while the rate of the ion exchange process was controlled by solute mass transfer through the film. However, Gao et al., 2010, found that the pseudo-second order kinetic model provided a better correlation for the adsorption of lactic acid by the Amberlite IRA-97 resin than the pseudo-first order kinetic model [112]. They reported that the lactic acid uptake process was mainly controlled by intraparticle diffusion at low concentrations (< 50 g lactic acid/L), but it was controlled by film diffusion at higher concentrations (> 70 g lactic acid/L), contrary than expected, and contrary to the findings of Zihao and Kefeng for the 2001 X 4 resin [101]. Gao et al. also investigated the thermodynamics of the ion exchange process. They observed that it is spontaneous and exothermic, and the values of ΔS^0 and ΔH^0 were estimated to be -0.82 $kJ.mol^{-1}.K^{-1}$ and -32.82 $kJ.mol^{-1}$, respectively.

When the capacities of the resins were investigated in batch conditions with synthetic and real fermentation broths, they were observed to decrease with respect to those obtained with aqueous lactic acid solutions [106,108,109,115-117]. The reason for this finding can be both the exchange of ions other than lactate (sulphate, phosphate, citrate, acetate, etc.) and the adsorption on the resin of extraneous components (nutrients, enzymes, etc.).

Table 3. Treatment of synthetic lactic acid (LA) solutions by means of anion exchangers

Resin type	Regeneration	Elution	Operating conditions	Results	Reference
VI-15 (Riedel-de-Haen), weak base	4% NaOH		30°C; 60 mg $LA.L^{-1}$; flow rate: $0.5 mL.min^{-1}$	Breakthrough point at pH 2.0: effluent volume of 20 mL; saturation capacity at pH 2.0: $0.645 g.g^{-1}$ dry resin. Breakthrough point and saturation capacity decrease with pH.	[100]
MWA-1 (Dow Chemical), moderate base	4% NaOH		30°C; 60 mg $LA.L^{-1}$; flow rate: $0.5 mL.min^{-1}$	Breakthrough point at pH 2.0: effluent volume of 17 mL; saturation capacity at pH 2.0: $0.420 g.g^{-1}$ dry resin. Breakthrough point and saturation capacity decrease with pH.	[100]
XUS 40196 (Dow Chemical), strong base	4% NaOH		30°C; 60 mg $LA.L^{-1}$; flow rate: $0.5 mL.min^{-1}$	Breakthrough point at pH 2.0: effluent volume of 175 mL; saturation capacity at pH 2.0: $0.355 g.g^{-1}$ dry resin.	[100]
201 x 4 (Nan Kai University), strong base	0.6 N NaOH	0.6 N NaOH	25°C; pH: 6.5	Degree of regeneration: 100 %	[101]
Amberlite IRA900 (Rohm and Hass), strong base	1 N NaOH	1 N HCl	45°C; pH: 4.85; 25 g $LA.L^{-1}$; flow rate: 3.5 $mL.min^{-1}$	Resin capacity: 0.273 $g.g^{-1}$ dry resin; 0.110 L of regenerant; 7.25 $g.L^{-1}$ lactic acid concentration; almost 100% recovery	[104]

Resin type	Regeneration	Elution	Operating conditions	Results	Reference
Amberlite IRA400 (Rohm and Hass), strong base	1 N NaOH	1 N HCl	45°C; pH: 4.85; 25 g $LA.L^{-1}$; flow rate: 3.5 $mL.min^{-1}$	Resin capacity: 0.279 $g.g^{-1}$ dry resin; 0.086 L of regenerant; 9.28 $g.L^{-1}$ lactic acid concentration; almost 100% recovery	[104]
Amberlite IRA96 (Rohm and Hass), weak base	1.0 N HCl	1.0 N HCl	25°C; pH: 4.85; 42 g $LA.L^{-1}$; flow rate: 3.5 $mL.min^{-1}$	Resin capacity: 0.377 $g.g^{-1}$ dry resin; 0.089 L of regenerant; 11.7 $g.L^{-1}$ lactic acid concentration; almost 100% recovery	[106]
Amberlite IRA67 (Rohm and Hass), weak base	1.0 N HCl	1.0 N HCl	25°C; pH: 4.85; 42 g $LA.L^{-1}$; flow rate: 3.5 $mL.min^{-1}$	Resin capacity: 0.374 $g.g^{-1}$ dry resin; 0.086 L regenerant; 13.3 $g.L^{-1}$ lactic acid concentration; almost 100% recovery	[106]
Indion 860 (Ion Exchange India), weak base	5.0% (w/v) NaOH	Metha nol	30°C; 13.5 g $LA.L^{-1}$; flow rate: 4.0$mL.min^{-1}$	Resin capacity: 0.347 $g.g^{-1}$ dry resin; 10 BV treated at breakpoint and 25 BV at bed exhaustion; 60 BV of regenerant for complete elution	[109]
Amberlite IRA92 (Rohm and Hass), weak base	1.0 N HCl	1.0 N HCl	43 g $LA.L^{-1}$; flow rate: 0.3 $mL.min^{-1}$; pH: 5.0	Lactic acid recovery: 98.67%	[110]

Table 4. Treatment of fermentation broths by means of anion exchangers

Resin type	Regeneration	Broth origin/ pretreatment	Operating conditions	Elution	Results	Reference
Amberlite IRA400 (Rohm and Hass), strong base	2.5 N NaOH	Synthetic media/ None	Coupled process pH: 6.0, T: 39°C	2.5 N HCl	Reduction of fermentation time by 69.4% and 5.32 fold increase of overall productivity (1.665 g. $L^{-1}.h^{-1}$) compared to discontinuous process	[95]
Amberlite IRA420 (Rohm and Hass), strong base	1 M Na_2CO_3	Synthetic media/ Microfiltration	Discontinuous 3 $BV.h^{-1}$, pH: 6.4	3 BV of 1 M Na_2CO_3	1st cycle: 69% of the LA exchanged by the resin; 15th cycle: 66% of the LA exchanged. Purity: 99%	[130]
Wolfatit SL 30 (Wolfen-Farben), strong base	0.1 N NaOH	Synthetic media/ None	Discontinuous/ Coupled; pH: 5.5 – 6.0 T: 37°C		Coupled process: adequate for pH control and increase in the overall productivity by 32.1% (0.6-0.7 $g.L^{-1}. h^{-1}$). Relatively pure solution of sodium lactate obtained	[99]

Resin type	Regeneration	Broth origin/ pretreatment	Operating conditions	Elution	Results	Reference
VI-15 (Riedel-de-Haen), weak base	3 BV 4% NaOH	Synthetic media/ None	Discontinuous pH: 2.9; 3 $BV.h^{-1}$	6.8 BV of methanol	Volume to reach saturation: 6.0 BV; Fraction of lactic acid in rinse effluent: 55%; LA recovered during elution: 100%; LA concentration: 31 $g.L^{-1}$	[131]
MWA-1 (Dow Chemical), moderate base	3 BV 4% NaOH	Synthetic media/ None	Discontinuous pH: 2.9; 3 $BV.h^{-1}$	1.5 BV of 5% NH_4OH	Volume to reach saturation: 1.5 BV; Fraction of lactic acid in rinse effluent: 51%; LA recovered during elution: 100%; LA concentration: 113 $g.L^{-1}$	[131]
Amberlite IRA 35 (Rohm and Hass), weak base	3 BV 4% NaOH	Synthetic media/ None	Discontinuous pH: 2.9; 3 $BV.h^{-1}$	5 BV of methanol	Volume to reach saturation: 5 BV; Fraction of lactic acid in rinse effluent: 48%; LA recovered during elution: 18%; LA concentration: 5 $g.L^{-1}$	[131]

Table 4. (Continued)

Resin type	Regeneration	Broth origin/ pretreatment	Operating conditions	Elution	Results	Reference
201 x 4 (Nan Kai University), strong base	0.6 N NaOH	Clarified synthetic media	Discontinuous, pH: 6.5, T: 25°C	0.6 N NaOH	The ion exchange efficiency was reduced when compared to LA model solutions.	[101]
Amberlite IRA400 (Rohm and Hass), strong base	1.0 N NaOH	Synthetic media/ Microfiltration	Coupled process	1.0 N HCl	Biomass production was 7-fold greater and LA productivity was 3-fold greater (7.75 $g.L^{-1}.h^{-1}$) when compared to discontinuous fermentation	[118]
D354 (Linan Resin Factory), weak base	NaOH	Synthetic media/ None	Coupled process. Immobilization of the microorganism	H_2SO_4 solutions	Utilization ratio of glucose: 99%; Productivity: 3-4 $g.L^{-1}.h^{-1}$. Both were much higher than those of discontinuous fermentation. Capacity: 0.5 $g.g^{-1}$ dry resin	[125,103]

Resin type	Regeneration	Broth origin/ pretreatment	Operating conditions	Elution	Results	Reference
Amberlite IRA400 (Rohm and Hass), strong base	1.5 N NaOH	Synthetic media/ None or pretreatment	Fluidized bed column, coupled process	1.5 N HCl/ 4.0 N HCl	The pretreatment proposed was able to increase the binding capacity by 120%. Capacity: 0.40 g $LA.g^{-1}$ dry resin. Elution: 90% of the bound product	[119,120, 121]
Amberlite IRA400 (Rohm and Hass), strong base	1.0 N NaOH	Pretreated Eucalyptus wood	Discontinuous, T: 45°C; pH: 4.85; 25 g $LA.L^{-1}$; flow rate: $3.5 mL.min^{-1}$	1.0 N HCl	Resin capacity: 0.140 $g.g^{-1}$ dry resin; 0.076 L regenerant; 5.08 $g.L^{-1}$ lactic acid concentration; almost 100% recovery	[104]
Amberlite IRA400 (Rohm and Hass), strong base	1.0 N NaOH	Pretreated Eucalyptus wood	Coupled, T: 45°C; pH: 4.85; 25 g $LA.L^{-1}$; flow rate: 3.5 $mL.min^{-1}$	1.0 N HCl	Overall process productivity: 1.18 g of $LA.L^{-1}.h^{-1}$ (31% higher than that of the discontinuous process). The pH was controlled within the optimum range (4.80 – 4.85).	[104]

Table 4. (Continued)

Resin type	Regeneration	Broth origin/ pretreatment	Operating conditions	Elution	Results	Reference
Amberlite IRA400 (Rohm and Hass), strong base	1.25 N NaOH	Synthetic media/ Pretreatment	Fluidized bed column, coupled process	4.0 N HCl	Best pretreatment: heating at 100°C and pH adjusted at 8.0. Recovery increased with column height. Total amount of LA recovered: 86%	[122]
Amberlite IRA400 (Rohm and Hass), strong base	1.0 N NaOH	Synthetic media/ None	Discontinuous; pH: 5.0 and 2.0, respectively	1.0 N H_2SO_4, 50% (v/v) methanol	Maximum capacity at pH 5.0: 0.1806 g LA.g^{-1} dry resin; Maximum capacity at pH 2.0: 0.10622 g LA.g^{-1} dry resin; Recovery: 86.21% at pH 5.0 and 89.20% at pH 2.0	[105]
Amberlite IRA96 (Rohm and Hass), weak base	1.0 N HCl	Pretreated Eucalyptus wood	Discontinuous, T: 25°C; pH: 4.85; flow rate: 3.5mL/min	1.0 N HCl	Resin capacity: 0.273 g.g^{-1} dry resin; 0.070 L regenerant; 16.0 g.L^{-1} lactic acid concentration; almost 100% recovery	[106]

Resin type	Regeneration	Broth origin/ pretreatment	Operating conditions	Elution	Results	Reference
Amberlite IRA400 (Rohm and Hass), strong base	4.0% (w/v) NaOH	Supplemented (yeast extract and lactose) whey/ None	Discontinuous. Agarose coated resin. T: 42°C; pH: 6.0	2.0 N HCl	Capacity: 0.076 $g.g^{-1}$ dry resin for uncoated resin and 0.072 $g.g^{-1}$ dry resin for coated resin. No adsorption of cells on the coated resin	[123]
Amberlite IRA400 (Rohm and Hass), strong base	4.0% (w/v) NaOH	Supplemented (yeast extract and lactose) whey/ None	Coupled. Agarose coated resin. T: 42°C; pH: 6.0	2.0 N HCl	Productivity increased by 40% when separation was integrated with fermentation. It was 4.1 $g.L^{-1}.h^{-1}$ for the uncoated resin and 3.1 $g.L^{-1}.h^{-1}$ for the coated one.	[123]
Amberlite IRA92 (Rohm and Hass), weak base	1.0 N H_2SO_4	Supplemented paper sludge/ None	Discontinuous. Ion exchange cromatography	1.0 N H_2SO_4	Optimum pH: 6.0. Under preliminary optimum conditions: yield: 82.6%; purity: 96.2%; productivity: 1.16 $gLA.g^{-1}$ resin day^{-1}	[108]

Table 4. (Continued)

Resin type	Regeneration	Broth origin/ pretreatment	Operating conditions	Elution	Results	Reference
Indion 860 (Ion Exchange India), weak base	5.0% (w/v) NaOH	Synthetic media/ None	Discontinuous	Methanol	Capacity: 0.280 $g.g^{-1}$ dry resin. 5 BV treated at breakpoint and 20 BV at bed exhaustion; 40 BV of regenerant for complete elution	[109]
Amberlite IRA400 (Rohm and Hass), strong base	2.5 N NaOH	Whey/ None	Coupled process	1.5 N NaOH	Optimum T: 37°C; Optimum pH: 6.0. Coupled process productivity (0.984 g. $L^{-1}.h^{-1}$) was 4.3 times higher and yield 0.7 times higher than those of the discontinuous process.	[124]
Amberlite IRA67 (Rohm and Hass), weak base	1.0 N HCl	Supplemented cassava bagasse/ None	Discontinuous	1.0 N HCl	Lactic acid recovery: 95.4%	[110]
Dowex Marathon WBA (Dow Chem.), weak base		Synthetic media/ None	Discontinuous. T: 30°C; pH: 6.0	1.0 N H2SO4 + H3PO4	Better performance at high flow rate; LA recovery yield: 76%; Purity: 90%.	[117]

Resin type	Regeneration	Broth origin/ pretreatment	Operating conditions	Elution	Results	Reference
Amberlite IRA400 (Rohm and Hass), strong base	1.0 N HCl, 1.0 N NaOH	Apple pomace + yeast extract/ Centrifugation	Discontinuous. T: 25°C; pH: 5.0	1.0 N HCl	Capacity: 0.222 $g.g^{-1}$dry resin. Quantitative recovery of LA.	[111]
Amberlite IRA96 (Rohm and Hass), weak base	1.0 N HCl, 1.0 N NaOH	Apple pomace + yeast extract/ Centrifugation	Discontinuous. T: 25°C; pH: 5.0	1.0 N HCl	Capacity: 0.360 $g.g^{-1}$ dry resin. Quantitative recovery of LA. Regeneration solution contained 55% of the LA fed, 30 g of $LA.L^{-1}$, LA purity: 92%	[111]
Amberlite IRA400 (Rohm and Hass), strong base	1.0 N HCl, 1.0 N NaOH	Apple pomace + corn steep liquor/ Centrifugation	Discontinuous. T: 25°C; pH: 5.0	1.0 N HCl	Capacity: 0.217 $g.g^{-1}$ dry resin. Quantitative recovery of LA	[111]
Amberlite IRA96 (Rohm and Hass), weak base	1.0 N HCl, 1.0 N NaOH	Apple pomace + corn steep liquor/ Centrifugation	Discontinuous. T: 25°C; pH: 5.0	1.0 N HCl	Capacity: 0.334 $g.g^{-1}$ dry resin. Quantitative recovery of LA	[111]
P4VP 8 (in situ prepared), weak base	1.0 N HCl	Clarified broth	Discontinuous	1.0 N HCl	Recovery yield: 95%; purity: 88%; capacity: 0.180 g $LA.g^{-1}$ dry resin	[113]

For example, Moldes et al., 2003, using pretreated Eucalyptus wood as simultaneous saccharification and fermentation media, determined that the resin capacities decreased by 27-35% with respect to those of the aqueous solutions [106]. Dave et al., 1997 and Dethe et al., 2006, using artificial solutions as feed streams, observed substantial glucose uptake by the Indion 860 resin, while phosphate and sulfate ions did not significantly interact with the resin [102,109]. On the other hand, Monteagudo and Aldavero, 1999, found that the capacity of the strong anionic exchange resin Amberlite IRA-420 increased with pH within the range 4.5 – 6.0 [115]. The ion exchange isotherms were reported to be of Langmuir form in most of the cases as well [106,115-117].

Tables 3 and 4 summarize the results obtained by different authors when the ion exchange process was performed by percolating the feed through a column packed with the resin. In some cases the ion exchange operation was performed separately from the fermentation (discontinuous process), while other authors tested an integrated operating mode where the ion exchange separation process was coupled with the fermentation (coupled process). Most of the works focused on investigating the capacities of the resins to exchange lactate ions and also on finding the most appropriate system to perform the elution of the lactic acid. However, little systematic research has been done on the effect of different operating conditions on the separation efficiency of lactic acid, and the recovery yield and purity of the lactic acid obtained was not indicated in most of the papers.

The selection of the best resin for lactic acid recovery should take into account not only the capacities of the resins for the substrate, but also the physical and mechanical properties of the resins, the easy of lactic acid desorption, the stability towards the load/regeneration cycles, the costs of the eluants used in the desorption and regeneration steps, and the cost of the treatment and disposal of the effluent solutions [100,106].

In the coupled processes, where the bioreactor was connected to the ion exchange column to continuously remove the lactic acid, the pH was controlled and maintained at the optimum value for the fermentation without the addition of alkali. Moreover, end-product inhibition was reduced. Nevertheless, the repeated recirculation of the eluted broth may be limited by the accumulation of inhibitory substances in the bioreactor [118]. Most of the authors used strong basic anion exchange resins in the OH^- form in order to favour pH control [95,96,99,115,116,118-124]. However, Lin et al., 1997, obtained satisfactory results with the weak-base resin D354 in the free base form. The pH value was maintained at about 3.5, which was adequate for the growth of the microorganism employed. They used several ion exchange columns in parallel. Thus, when one column was saturated, the broth was pumped into a new column, while the other was regenerated [125]. All the authors reported a great increase in overall lactic acid productivity, substrate conversion and biomass production and a significant reduction in fermentation time when the coupled process was compared to batch fermentation. Srivastava et al., 1999, proposed an empirical mathematical model, based on generalized Monod's growth model and Leudeking and Piret equation, to describe this coupled lactic acid fermentation process [126]. Lin et al., 1998, suggested as well a mathematical model able to describe the simultaneous process of lactic acid fermentation and separation by ion exchange [103].

In some of the works the broth was pretreated prior to the ion exchange step in order to remove cells and/or proteins that could adsorb on the resin. The techniques of microfiltration, ultrafiltration, filtration and centrifugation have been the most often considered [111,113,114,118,124,127-129]. In the membrane processes, the retentate was recycled back to the fermetor and the permeate was submitted to the subsequent ion exchange process. However, most of the authors percolated the broth directly through the column without any treatment. Some of them reported very little adsorption of the cells on the resin and very little effect of the presence of the resin on cell growth [95,105,107]. But Senthuram et al., 2004 observed that a significant amount of cells were able to adsorb onto the surface of the Amberlite IRA-400 resin [123]. In order prevent cell adsorption, they proposed shielding the resin with a thin layer of a non-ionic polymer (agarose). The coated resin did not show any significant cell adsorption, while the adsorption capacity was only 5% lower than that of the uncoted resin [109]. For the same purpose, Lin et al., 1997, immobilized the cells on a rotating disc contactor. Thus, the interaction of the cells with the fermentor elements (heat exchanger, baffles, etc) and with the ion exchange resin was avoided. They observed no visible growth of microorganisms in the column [125].

As can be observed when tables 3 and 4 are compared, the capacity of the resins is significantly reduced when fermentation broths are used as feed instead of aqueous lactic acid solutions, due to the competition of other species to adsorb on the column. For example, Models et al. reported that resin capacities decreased by 6-14% when fermentation broths obtained from the simultaneous saccharification and fermentation of pretreated Eucalyptus wood were processed by means of the Amberlite IRA-400 resin and by 27-35% when they were treated by the Amberlite IRA67 resin [106,116]. Most of the authors observed that the carbohydrate (mainly glucose) was also able to adsorb onto the resin surface to a certain extent. Glucose uptake was lower than 10% in most of the cases [102,103,111,130]. However, Cao et al., 2002, observed no adsorption of glucose on the Amberlite IRA-400 resin [105]. Vaccari et al., 1994, indicated that strong anions (such as sulfate) were fixed by the Amberlite IRA-420 resin; however, they were not later eluted together with the lactic acid when a solution of ammonium carbonate was used for this purpose [127]. Lihua et al., 1995, reported a negligible effect of inorganic salts on the ion exchange process [98] and most of the works showed insignificant adsorption of phosphates on the ion exchange resins tested [102,125].

A great number of different ion exchange resins were tested in the literature. In most of the eldest works, strong base anion exchangers in the OH^- form were used [95,99,101, 105,118-120, 122,123,130]. However, recent works demonstrated that the highest anion exchange capacities were obtained with weak-base anion exchange resins in the Cl^- form [106,110,113].

Some of the works focused on the selection of the best eluant to obtain the highest lactic acid recovery. If strong base exchangers are used, lactate can be eluted by means of strong desorbents such as NaOH. Sodium lactate is obtained and the resin is regenerated back to the OH^- form at the same time [101,124]. If lactic acid is required, an acid is normally used for lactate elution, being HCl the most frequent one [95,118-124]. Therefore, to recover 1 mol of lactic acid, 1 mol of base and acid are required for regeneration and elution, respectively. Thus, 1 mol of salt is generated, causing disposal problems [131]. The concentration of the acid solutions tested by different authors usually ranged from 1.0 to 4.0 N and the amount of lactic acid recovered during elution ranged from 86 to 100% of the adsorbed acid. However, Kulprathipanja and Oroskar, 1991, reported that the best desorbant was 0.002 – 0.02 N H_2SO_4 for strong as well as for weak anion exchangers [132]. Cao et al., 2002, tested various eluants to perform the recovery of lactic acid when artificial fermentation broths were treated by the Amberlite IRA-400 resin (solutions of H_2SO_4, methanol, NH_4OH, NaCl, water, HCl and mixtures of them). At pH 5.0 (above lactic acid pK_a) the best results in terms of lactic acid recovery (97.5% of the bound acid) were obtained with 2.5 M H_2SO_4. At pH 2.0 (below lactic acid pK_a) the elution can be performed with water (92.1% of the bound acid is recovered) [105]. John et al., 2008, reported that the best eluant was 1.0 N H_2SO_4 when lactic acid was recovered from cassava bagasse based fermented medium using the anion exchange resin Amberlite IRA-400 [110]. Vaccari et al., 1993, simultaneously regenerated back the Amberlite IRA-420 resin to the carbonate form and eluted the fixed lactic acid by means of ammonium carbonate solutions. After 3 bed volumes

(BV) of eluting agent, lactic acid was completely recovered in the form of ammonium salt. Afterwards, a partially concentrated ammonium lactate solution was treated with a strong cation exchange resin in the H^+ form to convert lactate into lactic acid. The authors reported a purity higher than 99% [130]. A similar approach was proposed by Monteagudo and Aldavero, 1999 and by Mantovani et al., 1992 [115,129].

Regarding the weak base resins, lactic acid is easily recovered by means of alcohols (ethanol and methanol), acetone, methyl acetate and even water. A free lactic acid product can be obtained after evaporating the eluent, which can be recycled without waste generation [131]. In the most recent works [106,110,111,113], the elution of lactic acid was performed with 1.0 N HCl to simultaneously obtain free lactic acid and regenerate the column to the Cl^- form. Separation of HCl from the final solution of lactic acid is easy due to its high volatility. Then it can be reused in the next elution step. Most of the authors reported quantitative recovery of lactic acid in the elution step [106,111]. Bookong et al., 2008, tested different eluants to perform the recovery of lactic acid. The Dowek Marathon WBA column was considered. The best results in terms of recovery were obtained with a mixture of 1.0 M sulfuric acid and 1.0 M phosphoric acid in a ratio of 30:70, however the purity was lower [117]. Some of the authors observed as well the elution of other components also adsorbed by the resin together with the lactic acid [100,131].

Córdoba et al, 1997 and 1999, Raya-Tonetti et al., 1999 and Sosa et al., 2001a and 2001b, explored the utilization of the strong anionic resin Amberlite IRA-400 in a fluidized bed column [116,119-122]. They observed that the capacity of the resin was low, but time of operation was much shorter than that obtained with the packed bed column. However, the authors reported that a pretreatment of the broth, consisting of the addition of 1.25 M NaOH, heating at 100°C during 15 min to eliminate the ammonium hydroxide, and finally pH adjustment to 8.0 with the same 1.25 M NaOH, was able to increase the binding capacity of the resin to a great extent (120 % increase). Lactic acid recovery was observed to increase with bed-height, operation with recirculation and decrease in lactic acid concentration. By maintaining an adequate ascensional velocity, higher than the biological solids terminal velocity, solids separation prior to the ion exchange step is avoided. Sosa et al., 2001b, proposed continuously well stirred tanks in series model to simulate the whole process.

A few authors investigated the effect of the operating variables on the recovery of lactic acid from fermentation broths by ion exchange. Temperature affects ion exchange equilibrium and kinetics. Temperature dependence of ion exchange equilibrium is related to the enthalpy change of the process. As ion exchange processes are, in general, slightly exothermic, an increase in temperature is expected to decrease the capacity of the resins, as it was observed by several authors [98,103,112]. However, the solute diffusion coefficients increase with temperature, therefore ion exchange velocity rises with this variable. Gluszcz et al., 2004, tested 18 different ion exchange resins to perform the recovery of lactic acid from lactic acid model solutions at pH values lower than 2. The saturation capacity of the resins was observed to decrease with temperature. However, the breakthrough capacity of the bed increased with temperature due to the increase of the diffusion coefficient within the sorbent particle [107]. Srivastava et al., 1992, observed that the binding ability of the Amberlite IRA-400 decreased with temperature as well. But lactic acid production in the fermentor increases with temperature within the range 37-45°C. The optimum temperature for the in situ separation of lactic acid from artificial fermentation broths was reported to be 39 °C [95]. Ataei and Forahani, 2008, found that the optimum temperature for the integrated process was 37°C, using the same resin and whey fermentation broths [124]. Several authors reported that the capacity of the strong anion exchange resins increased with pH due to the preferential adsorption of lactate ion over undissociated lactic acid on these resins [115]. The same findings were observed for the weak base anion exchange resins in the Cl^- or SO_4^{2-} forms [108,124]. However, the negative charge of the proteins in the fermentation broth increased as well with pH and they compete with lactic acid to adsorb on the resin. The optimum pH value in terms of recovery yield, purity and productivity was found to be 6.0 in most of the works [108,115,117,124]. On the contrary, Kulprathipanja and Oroskar, 1991, reported that when the strongly and weakly basic resins in sulfate form are used, the highest selectivity is reached at pH values below lactic acid pKa, as the nonionic form of lactic acid is preferentially adsorbed [132]. On the other hand, weak base resins in the free base form prefer undissociated lactic acid and their capacities decrease with pH, as it was previously explained. Boonkong et al., 2009, explored the effect of flow rate on the recovery of lactic acid from simulated fermentation broths by means of the anion exchange resin Dowex Maraton WBA. The authors found better adsorption performance at high flow rates, as narrower and steeper breakthrough curves were obtained [117].

From the results obtained in the aforementioned works, it seems that the integration of fermentation and ion exchange is a promising method to perform the production of lactic acid by fermentative processes, as it allows the simultaneous removal and recovery of lactic acid from the bioreactor and pH control within the optimum values. The utilization of MF or UF to remove the cells and/or proteins prior to the ion exchange is a good option. The best results in terms of kinetics, affinity for lactic acid, capacity and susceptibility to regeneration have been obtained with weak anion exchangers in the Cl^- form, which can be eluted and regenerated by means of HCl solutions. However, in most of the works, the recovery yield and purity of the lactic acid obtained were not good enough and further purification steps would be required. The recovery yields, purities and productivities obtained in the different works vary to a great extent, depending on the resin type, fermentation media and operating conditions tested, as can be observed from table 4. Some of the proposed processed have been patented [114,129,132,133].

Recovery of Lactic Acid by Cation and Anion Exchangers

A novel approach proposed by the authors of this chapter to obtain purified lactic acid solutions from fermentation broths consists of the removal of undesired cations and anions from the broths by means of cation and anion exchangers while the lactic acid remains in solution [9]. The process was tested with artificial solutions and with real fermentation broths obtained using sweet whey supplemented with yeast extract as fermentation media. The fermentation broth was clarified by means of ultrafiltration prior to the ion exchange process. Thus, microorganisms, proteins and lipids were entirely removed from the broth. The optimum conditions for the fermentation step were observed to be 42°C and pH 5.6.

The cation and anion exchange resins selected were, respectively, Lewatit S2568H (strongly acidic) and Lewatit S3428 (weakly basic). The regeneration of the cation exchanger was performed with 2.5 BV of 1 M HCl at room temperature and the anion exchanger was regenerated with 6 BV of 0.75 M NaOH at room temperature as well. A detailed description of the operating procedure and analytical methods can be found elsewhere [9].

The ion exchange process consisted of two steps:

1) The clarified fermentation broth was percolated through the cation exchanger to eliminate the cations present in the broth. At the same time the pH was reduced and lactate ions were converted into lactic acid.

2) The effluent from the cation exchanger was passed through the anion exchanger to remove undesired anions from the broth, while most of the lactic acid is expected to remain in solution. The selectivity of the anion exchanger was tested with artificial solutions and real fermentation broths.

The effect of the operating conditions (temperature, flow rate, pH) on both ion exchange processes was investigated. The effect of temperature and flow rate on the performance of the cation exchanger was not significant, though retention slightly increased when decreasing the flow rate and temperature within the range considered (13-33 $BV.h^{-1}$ and 25-40°C). The cation exchanger was able to produce 4 BV of acidified (pH of 1.5) and cation free broth regardless the temperature and flow rate considered.

Regarding the anion exchanger, in order to be able to separate lactic acid from the other acids present in the broth, the resin should show high affinity for those acids and low affinity for lactate ion. The affinity of the resin for the main anions in the broth follows the order chloride > phosphate > lactate. Thus, their separation is possible. It was demonstrated that the capacity and the selectivity of the resin for phosphate and chloride over lactate was much higher at very acidic pH values (about 1.2) than at the pH value of the fermentation broth (5.6).

The highest selectivity of the resin was obtained at the temperature of the fermentation broth (i.e. 40°C). Nevertheless, the effect of temperature was not very significant either. Increasing the flow rate resulted in more spread out breakthrough curves. Therefore, high flow rates are not recommended as they result in a selectivity reduction and lactate loses increase.

At the optimum conditions selected, the resin hardly retained the lactate ion, while phosphate and chloride were highly retained. Thus, 25 BV of purified lactic acid was obtained with good selectivity. The purity of the product was higher than 99% and lactate loses were estimated to be lower than 15%.

The economic profitability of the process has been demonstrated as well [97]. The results obtained are better than those previously reported; however, the main drawback of the process is the need of chemicals to regenerate the resins. Afterwards, highly acidic and basic solutions are obtained, which also contain sodium, potassium, magnesium, calcium and phosphates and chlorides, and should be therefore treated.

REFERENCES

[1] Datta, R. and Henry, M., *Journal of Chemical Technology and Biotechnology*, 81 1119-1129 (2006).

[2] Vijayakumar, J., Aravindan, R. and Viruthagiri, T., *Chemical and Biochemical Engineering Quarterly*, 22 (2) 245-264 (2008).

[3] Huang, C., Xu, T., Zhang, Y., Xue, Y. and Chen, G., *Journal of Membrane Science*, 288 1-12 (2007).

[4] Gonçalves, L. M. D., Xavier, A. M. R. B. and Almeida, J. S., *Enzyme and Microbial Technology*, 13 311-319 (1991).

[5] Ohara, I., Hiyama, K. and Yoshida, T., *Applied Microbiology and Biotechnology*, 37 544-548 (1992).

[6] Marinova, M., Kyuchoukov, G., Albert, J., Molinier, J. and Malmary, G., *Separation and Purification Technology*, 37 199-207 (2004).

[7] Yi, S.S., Lu, Y.C. and Luo, G.S., *Separation and Purification Technology*, 60 308-314 (2008).

[8] Eyal, A.M., Starr, J.N., Fisher, R., Hazan, B., Canari, R. and Witze, D.R., US Patent 6,229,046 (2001).

[9] González, M.I., Alvarez, S., Riera, F.A. and Alvarez R., *Industrial & Engineering Chemistry Research*, 45 3243-3247 (2006).

[10] Banniel, A.M., Eyal, A.M., Mizahari, J., Hazan, B and Fisher, R., US Patent, 6,187,951 (2001).

[11] Madzingaidzo, L., Danner, H. and Braun, R., *Journal of Biotechnology*, 96 223-239 (2002).

[12] Polom, E. and Szaniawska, D., *Desalination*, 198 208-214 (2006).

[13] González, M.I., Alvarez, S., Riera, F.A. and Alvarez R., *Desalination*, 228 84-96 (2008).

[14] Wasewar, K.L., *Chemical and Biochemical Engineering Quarterly*, 19(2) 159-172 (2005).

[15] Joglekar, H.G., Rahman, I., Babu, S., Kulkarni, B.D. and Joshi, A., *Separation and Purification Technology*, 52 1-17 (2006).

[16] Hu, W.S.and Sirkar, K.K., *Membrane Handbook*, Van Nostrand Reinhold, N.Y., (1992).

[17] Mulder, M. (ed.), *Basic Principles of Membrane Technology*, Kluwer Academic, Dordretht, (1996).

[18] Cheryan, M., *Microfiltration and Ultrafiltration Handbook*. Technomic Publishing Co., Lancaster, Pennsylvania (1998).

[19] Drioli, E. and Giorno, L. (eds.), *Comprehensive Membrane Science and Engineering*, Elsevier, Kidlington (2010).

[20] Cui, Z.F. and Muralidhara, H.S. (eds.), *Membrane Technology*, Elsevier, Amsterdam (2010).

[21] Cheryan, M., *Ultrafiltration Handbook*, Technomic Publishing Co., Lancaster, Pennsylvania (1986).

[22] Taniguchi, M., Kotani, N. and Kobayashi, T., *Journal of Fermentation Technol*ogy, 65(2) 179-184 (1987).

[23] Crespo, J.P.S.G., Xavier, A.M.R.B., Bareto, M.T.O., Gonçalves, L.M.D., Almeida, J.S. and Carrondo, M.J.T., *Chemical Engineering Science*, 47 204-214 (1992).

[24] Xavier, A.M.R.B., Gonçalves, L.M.D., Moreira, J.L.and Carrondo, M.J.T., *Biotechnology and Bioengineering*, 45(4) 320-327 (1995).

[25] Moueddeb, H., Sanchez, J., Bardot, C. and Fick, M., *Journal of Membrane Science*, 114 59-71 (1996).

[26] Fu, W. and Mathew, A.P., *Biochemical Engineering Journal*, 3 163-170 (1999).

[27] Nishiwaki, A.and Dunn, I.J., *Bioprocess Engineering*, 21 299-305 (1999).

[28] Torang, A., Jonsson, A.S.and Zacchi, G., *Applied Biochemistry and Biotechnology*, 76 144-147 (1999).

[29] Milcent, S.and Carrère, H., *Separation and Purification Technology*, 22 393-341 (2001).

[30] Kwon, S., Yoo, I.K., Lee, W.G., Chang, H.N. and Chang, Y.K., *Biotechnology and Bioengineering*, 73 25-34 (2001).

[31] Giorno, L., Chojnacka, K., Donato, L. and Drioli, E., *Industrial & Engineering Chemistry Research*, 41 433-440 (2002).

[32] Levente, L. D. and Puzanov, T., *International Journal of Applied Science and Engineering*, 3(1) 19-25 (2005).

[33] Xu, G., Chu, J., Wang, Y., Zhuang, Y., Zhang, S. and Peng, H., *Process Biochemistry*, 41 2458-2463 (2006).

[34] Pall, P., Sikder, J., Roy, S. and Giorno, L., *Chemical Engineering and Processing*, 48 1549-1559 (2009).

[35] Jeantet, R., Maubois, J.L.and Boyaval, P., *Enzyme and Microbial Technology*, 19 614-619 (1996).

[36] Cuartas-Uribe, B., Alcaina, M.I., Soriano, E., Mendoza, J.A., Iborra, M.I. and Lora, J., *Desalination*, 241 244-255 (2009).

[37] Lo Presti, S., Moresi, M., *Journal of Food Engineering*, 44 109-117 (2000).

[38] Moresi, M., Ceccantoni, B. and Lo Presti, S., *Journal of Membrane Science*, 209 405-420 (2002).

[39] Cavaco, A.I., Szymczyk, A., Fievet, P.and Brites, A.M., *Journal of Membrane Science*, 332 320-330 (2008).
[40] Boussu, K., Vandecastleele, B. and Van der Bruggen, B., *Journal of Membrane Science*, 310 51-65 (2008).
[41] Pontalier, P., Ismail, A. and Ghoul, M., *Separation and Purification Technology*, 12 175-181 (1997).
[42] Hagmeyer, G.and Gimbel, R., *Desalination*, 117 247-256 (1998).
[43] Schaep, J., Vandecasteele, C., Mohammad, W. A.and Bowen, W.R., *Separation and Purification Technology*, 22-23 169-179 (2001).
[44] Bowen, W.R., Welfoot, J.S., *Chemical Engineering Science,* 57 1121-1137 (2002).
[45] Tanninen, J. and Nyström, M., *Desalination,* 147 295-299 (2002).
[46] Straatsma, J., Bargeman, G., van der Horst H.C. and Wesselingh, J.A., *Journal of Membrane Science*, 198 273-284 (2002).
[47] Van der Bruggen, B. and Vandecasteele, C., *Water Research*, 36 1360-1368 (2002).
[48] Van der Bruggen, B., Koninckx, A. and Vandecasteele, C., *Water Research*, 38 1347-1353 (2004).
[49] Labbez, C., Fievet, P., Szymezyck, A., Vidonne, A., Foisy, A. and Pagetti, J., *Separation and Purification Technology*, 30 47-55 (2003).
[50] Bargeman, G., Vollenbroek, J.M., Straatsma, J., Schroen, C.G.P.H. and Boom, R.M., *Journal of Membrane Science*, 247 11-20 (2005).
[51] Tanninen, J., Mänttari, M.,and Nystrom, M., *Journal of Membrane Science*, 294 201-212 (2007).
[52] Timmer, J.M.K., van der Horst, H.C.and Robbertsen, T., *Journal of Membrane Science*, 85 205-216 (1993).
[53] Timmer, J.M.K., van der Horst, H.C. and Robbertsen, T., *Journal of Membrane Science*, 92 185-197 (1994).
[54] Liew, M.K.H., Tanaka, S. and Morita, M., *Desalination*, 101 269-277 (1995).
[55] Bouchoux, A., Roux-de Balmann, H.and Lutin, F., *Journal of Membrane Science*, 258 123-132 (2005).
[56] Duke, M., Lim, A., Castro da Luz, S. and Nielsen, L., *Food and Bioproducts Processing*, 86 290-295 (2008).
[57] Li, Y., Shahbazi, A., Williams, K. and Wan, C., *Applied Biochemistry and Biotechnology.*, 147 1-9 (2008).
[58] Zhuo, K., Wang, J., Yue, Y. and Wang, H., *Carbohydrate Research*, 328 383-391 (2000).

[59] Umpuch, C., Galier, S., Kanchanatawee, S. and Roux-de Balmann, H., *Process Biochemistry*, 45 1763-1768 (2010).
[60] Schaep, J., Vandecasteele, C., Mohammad, A.W. and Bowen, W.R., *Separation and Purification Technology*, 34 3009-3030 (1999).
[61] Ramadan, Y., Pátzay, G. and Szabó, G.T., *Desalination*, 256 54-57 (2010).
[62] Van der Bruggen, B., Schaep, J., Wilms, D. and Vandecasteele, C., *Journal of Membrane Science*, 156 29-41 (1999).
[63] Bouranene, S., Szymczyk, A., Fievet, P. and Vidonne, A., *Journal of Membrane Science*, 290 216-221 (2007).
[64] Li, Y., Shahbazi, A. and Kadzere, C., *Journal of Food Engineering*, 75 574-580 (2006).
[65] Martin, A., Martinez, F., Malfeito, J., Palacio, P., Prádanos, P. and Hernández, A., *Journal of Membrane Science*, 213 225-230 (2003).
[66] Cavaço, A.I., Brites, A.M., Costa, M.C. and Cardoso, J.P., *Chemical Engineering Science*, 61 2418-2427 (2006).
[67] Xu, Y. and Lebrun, R.E., *Journal of Membrane Science*, 158 93-104 (1999).
[68] Mandale, S. and Jones, M., *Desalination,* 219 262-271 (2008).
[69] Childress, A. E. and Elimelech, M., *Environonmental Science and Technology*, 34 3170-3716 (2000).
[70] Susanto, H. and Ulbricht, M., *Journal of Membrane Science*, 266 132-142 (2005).
[71] Takahashi, H., Ohba, K. and Kikuchi, K.I., *Biochemical Engineering Journal*, 16 311-315 (2003).
[72] Wee, Y.J., Yun, J.S., Lee, Y.Y., Zeng, A.P. and Ryu, H.W., *Journal of Bioscience and Bioengineering*, 99 104-108 (2005).
[73] Huang, C.H. and Xu, T.W., *Environonmental Science and Technology*, 40 5233-5243 (2006).
[74] Strathmann, H., *Ion-exchange Membrane Separation Processes*, Elsevier, Amsterdam, 2004.
[75] Xu, T.W., *Journal of Membrane Science*, 263 1-29 (2005).
[76] Yi, S.S., Lu, Y.C. and Luo, G.S., *Journal of Membrane Science*, 255 57-65 (2005).
[77] Nagarale, R.K., Gohil, G.S. and Shahi, V.K., *Advances in Colloid and Interface Science*, 119 97-130 (2006).
[78] Cowan, D.A. and Brown, J.H., *Industrial and Engineering Chemistry,* 51 1445-1555 (1959).

[79] Widiasa, I.N., Sutrisna, P.D. and Wenten, I.G., *Separation and Purification Technology*, 39 89-97 (2004).
[80] Thang, V.H., Koschuh, W., Kulbe, K.D. and Novalin, S., *Journal of Membrane Science*, 249 173-182 (2005).
[81] Pourcelly, G. and Gavach, C., *Electrodialysis Water Splitting-Application of Electrodialysis with Bipolar Membranes*, in: A.J.B. Kemperman (ed.) Handbook on Bipolar Membrane Technology, Twente University Press, Enschede, The Netherlands (2000).
[82] Czytko, M., Ishii, K. and Kawai, K., *Chemie Ingenieur Technik,* 59 952-954 (1987).
[83] Danner, H., Madzingaidzo, L., Holzer, M., Mayrhuber, L. and Braun, R., *Bioresource Technology*., 75 181-187 (2000).
[84] Lee, E.G., Moon, S.H., Chang, Y.K., Yoo, I.K. and Chang, H.N., *Journal of Membrane Science*, 145 53-66 (1998).
[85] Bailly, M., Roux-de Balmanna, Aimar, P., Lutin, F. and Cheryan, M., *Journal of Membrane Science*, 191 129-142 (2001).
[86] Hábová, V., Melzoch, K., Rychtera, M. and Sekavará, B., *Desalination*, 163 363-372 (2004).
[87] Lee, K., *Bioresource Technology*, 96 1505-1510 (2005).
[88] Kim, Y.H. and Moon, S-H., *Journal of Chemical Technology and Biotechnology*, 76 169-178 (2001).
[89] Li, H., Mustacchi, R., Knowles, C.J., Skibar, W., Sunderland, G., Dalrymple, I. and Jackman, S.A., *Tetrahedron*, 60 655-661 (2004).
[90] Hirata, M., Gao. M-T., Toorisaka, E., Takanashi, H. and Hano, T., *Biochemical Engineering Journal*, 25 159-163 (2005).
[91] Resbeut, S., Pourcelly, G., Sandeux, R. and Gavach, C., *Desalination,* 120 235-245 (1998).
[92] Montiel, V., García, V., González, J., Carmona F. and Aldaz, A., *Journal of Membrane Science*, 140 243-250 (1998).
[93] Lee, E.G., *Recovery of Lactic Acid from Fermentation Broth Using Electrodialysis*, PhD. Thesis, Dept. of Chemical Engineering, Korea Advanced Institute of Science and Technology, Korea (1998).
[94] Helfferich, F., *Ion Exchange*, Dover Publications, New York (1995).
[95] Srivastava A., Roychoudhury P. and Sahai V., *Biotechnology and Bioengineering,* 39 607-613 (1992).
[96] Sun,Y; Li, Y.-L.; Bai, S. and Hu, Z.-D., *Industrial and Engineering Chemistry Research,* 38(9) 3290–3295 (1999).
[97] González, M.I.; Álvarez, S.; Riera, F.A. and Álvarez, R., *Journal of Food Engineering*, 80 553-561 (2007).

[98] Lihua, X.; Shandong L.; Zhongming, Z.; Xinhua, D. and Peilin, C., *Chinese Journal of Chemical Engineering,* 3(2) 82-87 (1995).
[99] Beschkov, V.; Peeva, L. and Valchanova, B., *Hungarian Journal of Industrial Chemistry Veszprém,* 23 135-139 (1995).
[100] Evangelista, R.L.; Mangold, A.J. and Nikolov, Z., *Applied Biochemistry and Biotechnology,* 45/46 131-144 (1994).
[101] Zihao, W. and Kefeng, Z., *Biotechnology and Bioengineering,* 47 1-7 (1995).
[102] Dave, S.M., Patil, S.S. and Suresh, A.K., *Separation Science and Technology,* 32(7) 1273-1294 (1997).
[103] Lin, J.P., Ruan, S.D. and Cen, P.L., *Chemical Engineering Communications,* 168 59-79 (1998).
[104] Moldes, A.B., Alonso, J.L. and Parajó, J.C., *Applied Biochemistry and Biotechnology*, 95 69-81 (2001).
[105] Cao, X, Yun, H.S. and Koo, Y-M., *Biochemical Engineering Journal*, 11 189-196 (2002).
[106] Moldes, A.B., Alonso, J.L. and Parajó, J.C., *Bioprocess and Biosystems Engineering*, 25 357-363 (2003).
[107] Gluszcz, P., Jamroz, T., Sencio, B. and Ledakowicz, S., *Bioprocess and Biosystems Engineering,* 26 185-190 (2004).
[108] Tong, W.Y., Fu, X.Y., Lee, S.M., Yu, J., Liu, J., Wei, D.Z. and Koo, Y.M., *Biochemical Engineering Journal*, 18 89-96 (2004).
[109] Dethe, M.J., Marathe, K.V. and Gaikar, V.G., *Separation Science and Technology,* 41 2947-2971 (2006).
[110] John, R.P., Nampoothiri, K.M. and Pandey, A., *Brazilian Archives of Biology and Technology,* 51(6) 1241-1248 (2008).
[111] Gullón, B., Alonso, J.L. and Parajó, J.C., *Industrial and Engineering Chemistry Research,* 49 3741-3750 (2010).
[112] Gao, Q., Liu, F., Zhang, T., Zhang, J., Jia, S., Yu, C., Jiang, K. and Gao, N., *Plosone*, 5(11) e13948 (2010).
[113] Zhou, J.; Bi, W. and Row, K.H., *Journal of Applied Polymer Science,* 120 2673-2677 (2011).
[114] Hofman, M., European Patent 0 265 409 A1 (1988).
[115] Monteagudo, J.M. and Aldavero, M., *Journal of Chemical Technology and Biotechnology,* 74 627-634 (1999).
[116] Sosa, A.V., Ochoa, J. and Perotti, N.I., *Bioseparation,* 9 283-289 (2001).
[117] Boonkong, V., Sangvanich, P., Petsom, A. and Thongchul, N., *Chemical Engineering and Technology,* 32(10) 1542-1549 (2009).

[118] Ye, K.; Jin, S. and Shimizu, K., *Journal of Chemical Technology and Biotechnology,* 66 223-226 (1996).
[119] Córdoba, P.R., Ragout, A.L., Siñeriz F. and Perotti, N.I., *Biotechnology Techniques,* 8 629-634 (1997).
[120] Córdoba, P., Raya Tonetti, G., Siñeriz F. and Perotti, N.I., *Latin American Applied Research,* 29 31-34 (1999).
[121] Raya Tonetti, G.; Córdoba, P.; Bruno-Bárcena, J.; Siñeriz F. and Perotti, N., *Biotechnology Techniques,* 13 201-205 (1999).
[122] Sosa, A.V., Córdoba, P.R. and Perotti, N.I., *Biotechnology Progress,* 77 1079-1083 (2001).
[123] Senthuran, A., Senthuran, V., Hatti-Kaul, R. and Mattiasson, B., *Applied Microbiology and Biotechnology,* 65 658-663 (2004).
[124] Ataei, S.A. and Vasheghani-Farahani, E., *Journal of Industrial Microbiology and Biotechnology,* 35 1229-1233 (2008).
[125] Lin, J., Bo, C., Jianping, W. and Peilin, C., *Chinese Journal of Chemical Engineering,* 5(1) 49-55 (1997).
[126] Srivastava A., Yunus, R. and Roychoudhury, P.K., *Artificial Cells, Blood Substitutes and Immobilization Biotechnology,* 27 403-410 (1999).
[127] Vaccari, G., Dosi, E., Campi, A.L. and Mantovani, G., *Zuckerindustrie,* 119(3) 225-229 (1994).
[128] Stroia, I., Begea, M., Boca, E., Baron, E., Baldea, G.H., Musu, S., Mihai, C. and Vamanu, E., *Journal of Environmental Protection and Ecology,* 8 401-409 (2007).
[129] Mantovani, G., Vaccari, G. and Campi, A.L., European Patent, 0 517 242 A2 (1992).
[130] Vaccari, G., González-Vara, A.R., Campi, A.L., Dosi, E.; Brigidi, P. and Matteuzzi, D., *Applied Microbiology and Biotechnology,* 40 23-27 (1993).
[131] Evangelista, R.L. and Nikolov, Z., *Applied Biochemistry and Biotechnology,* 57/58 471-480 (1996).
[132] Kulprathipanja, S. and Oroskar, A.R., United States Patent 5 068 418 (1991).
[133] Sanovations, B.V., European Patent, 2 239 333 A1 (2010).

In: Lactic Acid
Editors: A. Migallón and R. Ruseckaite

ISBN: 978-1-62081-643-1

Chapter 2

LACTIC ACID PRODUCTION FROM RENEWABLE RESOURCES

Luciana Fontes Coelho, Marcela Piassi Bernardo, Paola Monteiro de Oliveira, Cristian Jacques Bolner de Lima and Jonas Contiero*

UNESP – Univ. Estadual Paulista, Biological Sciences Institute, Department of Biochemistry and Microbiology, Rio Claro, SP, Brazil

ABSTRACT

Lactic acid has been used in pharmaceutical, cosmetic, chemical and food industries. It can be used as a raw material in biodegradable plastics, such as polylactic acid (PLA) as well as to improve the physical properties in the production of garbage bags, agricultural plastic sheeting, and food packaging. It can also be used in sutures and surgical implants due to its biocompatible and bioabsorbable characteristics. Lactic acid is industrially produced either through chemical synthesis or microbial fermentation.

The advantage of the biological method is that an optically pure lactic acid can be obtained by choosing a strain of lactic acid bacteria, whereas chemical synthesis always results in a racemic mixture of lactic acid. The optical purity of lactic acid is very important to the physical

* Corresponding Author: Jonas Contiero, Department of Biochemistry and Microbiology, Univ. Estadual Paulista - UNESP; CEP 13506 – 900 Rio Claro – SP, Brazil; Tel.: +55 19 35264180 Fax: +55 19 35264176 e-mail: jconti@rc.unesp.br.

properties of PLA and obtaining a more stable crystalline polymer than that achieved with a racemic lactic acid. Therefore, raw materials for the industrial production of lactic acid need to have characteristics such as low cost, low levels of contaminants, rapid fermentation, and year-round availability. The cost of raw materials represents 68% of the total cost of lactic acid production. A number of industrial byproducts or wastes have been evaluated as substrates for lactic acid production with the aim of decreasing the cost of the process, such as sugarcane, molasses, cassava wastewater, and whey as carbon sources and corn steep liquor (CSL) and yeast autolysate as nitrogen sources. In this chapter, it will be discussed the fermentation technology to lactic acid production that has been used and exploited by using the cheaply available source materials.

Keywords: Lactic acid; poly(lactic acid); lactic acid bacteria; fermentation; biodegradable polymer

INTRODUCTION

Lactic acid has been used as a raw material in biodegradable plastics, such as poly(lactic acid) (PLA), and can be used to improve the physical properties in the production of garbage bags, agricultural plastic sheeting, and food packaging [1]. It can also be used in sutures and surgical implants due to its biocompatible and bioabsorbable characteristics [2]

Lactic acid is industrially produced either through chemical synthesis or microbial fermentation. The advantage of the biological method is that an optically pure lactic acid can be obtained by choosing a strain of lactic acid bacteria, whereas chemical synthesis always results in a racemic mixture of lactic acid [3]. The optical purity of lactic acid is very important to the physical properties of PLA and obtaining a more stable crystalline polymer than that achieved with a racemic lactic acid [4,5]. Polymers made with L(+)-lactic acid have a melting point of 175 °C. However, this melting point can be increased by adding D(−)-lactic acid, thereby producing a stereocomplex with a melting point of around 230 °C [6,7].

Lactic bacteria have complex nutritional requirements due to their limited ability to biosynthesize B vitamins and amino acids [8]. The use of pure sugars and pure complex nitrogen sources is expensive. Therefore, raw materials for the industrial production of lactic acid need to have characteristics such as low cost, low levels of contaminants, rapid fermentation, and year-round availability [3].

According to Tejayadi and Cheryan [9], the cost of raw materials represents 68% of the total cost of lactic acid production. A number of industrial bioproducts or wastes have been evaluated as substrates for lactic acid production with the aim of decreasing the cost of the process, such as sugarcane [10], molasses [11], cassava wastewater [12], and whey [13] as carbon sources and corn steep liquor (CSL) [14] and yeast autolysate [15] as nitrogen sources. Sugarcane juice, sugarcane molasses and whey are renewable, abundant, and cheap sources of carbon. They were also proven to be an economically feasible raw material for industrial production of lactic acid once they have enough nutrients necessary for the growth of lactic acid bacteria. Furthermore, sugarcane molasses and sugarcane juice do not require treatment to convert starch to fermentable sugars as the sugar content is almost all in the form of sucrose [10] or in the form of lactose in the case of whey. This way, it is possible to reduce the cost of culture medium by using these substrates in order to increase economic viability.

These substrates are renewable, abundant, and cheap sources of carbon and they have enough nutrients necessary for the growth of lactic acid bacteria. Although agroindustrial lignocellulosic waste require a pretreatment by acid or enzymatic hydrolysis to provide the fermentable sugars, it is important to highlight the importance of these raw materials, mainly because of the low cost and high availability. Thus, some studies have been conducted with the objective of developing viable technologies that improve the fermentation of lactic acid from sugarcane bagasse, and the fermentation of pentoses from hemicellulose.

Utilization of renewable resources is an alternative to reduce environmental impact, such as global warming, as well as to reduce the costs of production. The use of biomass is a promising possibility, especially for the countries that produces abundant, diverse and cheap raw materials. The conversion of biomass into a product with higher added value that can replace petroleum derivatives, through the production of green polymers has great potential that has not been well explored yet.

Lactic Acid and Polylactic Acid

The lactic acid ($CH_3CHOHCOOH$) is an organic acid with an asymmetric carbon atom with four different groups attached to it: -COOH,-H,-OH and CH_3. There are two enantiomerics forms (Figure 1). Lactic acid is widely used in textile, pharmaceutical, chemical and food industries.

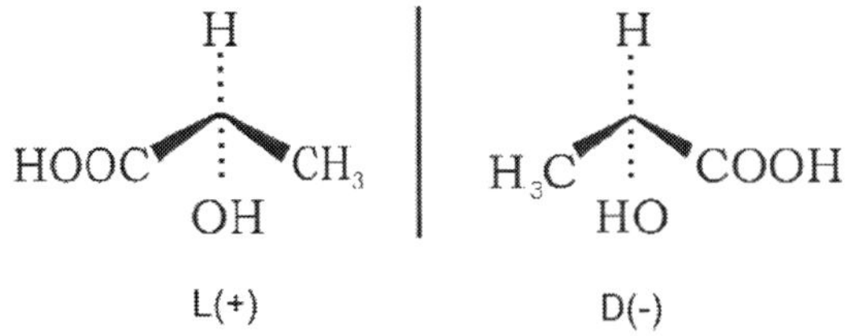

Figure 1. Stereoisomers of lactic acid.

The use of lactic acid in the pharmaceutical industry is related to the production of cosmetics. Calcium lactate can be used as therapy for calcium deficiency and it can also act as an agent for the disposal of dental caries [16]. In the food industry, lactic acid is used as an additive, acidulant, flavoring and emulsifier, and inhibiting the bacteria sporulation in processed food. In addition, chemical industry employed it in the organic solvents, 1,2-propanediol and acrylic acid production [17]. The poly(lactic acid) has been considered one of the most promising biodegradable plastics. The high chemical resistance of this compound is advantageous for the manufacture of fibers and films, while the heat resistance is favorable for production of diverse utensils such as cups, trays [18], cutlery, plates, bags, bags fertilizer, soda bottles, grocery bags, etc [19]. The polymer produced from D-(-) and L-(+) lactic acid has shown physical and mechanical properties comparable to that of plastics produced from fossil energy reserves, but with high rates of biodegradability. It can be totally degraded in the environment under aerobic or anaerobic conditions [20]. In the synthesis of PLA, two lactic acid molecules undergo a process of esterification and catalytic cyclization to a cyclic ester lactide (Figure 2). This process of dimerization generates water, which can be separated prior to polymerization due to a significant decrease in polarity. High molecular weight PLA is produced from the ester lactide by ring opening polymerization by using a catalyst (stannous octoate). This mechanism does not generate additional water molecules, and therefore it is possible to obtain a wide range of molecular weights [21].

The enantiomeric purity of lactic acid is a crucial factor in the physical properties of PLA. Polymerization of a racemic mixture of L-(+) or D-(-) lactic acid usually leads to the synthesis of PDLLA, which is amorphous. The degree of crystallinity, and many other important properties are controlled by the ratio of the enantiomers D- and L- [22]. Several distinct forms of polymers can exist: PLLA is the product resulting from the polymerization of L-(+) lactic acid.

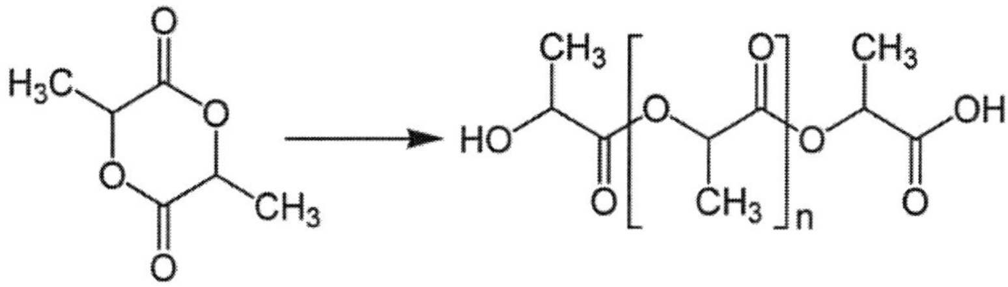

Figure 2. Synthesis of poly-lactic acid.

PLLA has a crystallinity of 37%, and a melting temperature between 173 and 178 °C. The melting temperature of PLLA can be increased up to 50 °C and its heat deflection temperature can be increased by approximately 60 °C up to 190 °C by physical mixing with the polymer PDLA, which can form a highly regular stereocomplex with higher crystallinity.

The temperature stability is maximized when a mixture 50:50 is used, but even at low concentrations from 30 to 10% of PDLA, there is still a substantial improvement. PDLA acts as a nucleating agent, thereby increasing the crystallization rate. However, due to the higher crystallinity of PDLA, biodegradation is slower than for the PLA [21].

Lactic acid can be achieved either by the action of fermentative bacteria, fungi and yeasts and by chemical synthesis. However, the fermentation processes are more advantageous because they are more economical, not requiring the use of high pressures and temperature, metal catalysts, and energy consumption. In addition, the cost of raw materials is low, because it is possible to use "stereoselective" renewable raw materials, to produce preferentially an isomer [23].

Furthermore, chemical synthesis of lactic acid always produces a racemic mixture. Approximately 90% of total lactic acid produced worldwide is by fermentation using bacteria [24]. Lactic acid bacteria are very picky about their growth conditions.

According to Buchta [25], sugars represent the best carbon sources for these bacteria, there also need of nitrogen, vitamins and minerals for the proper performance of lactic acid fermentation. Several by-products and raw materials of agro-industry and food industry have been used to produce lactic acid, because of their high availability and low cost, such as bagasse, molasses and sugarcane juice, whey, besides of the CSL (corn steep liquor) and of yeast autolysate.

During the fermentation of sugar, depending on the species of bacteria, they can produce only L-(+) lactic acid, only D-(-) lactic acid, approximately equal amounts of both isomers [26-28]. Most of the lactic acid bacteria produce both enantiomeric forms of lactic acid. The enzymatic reduction of pyruvate to L-(+) is formed by the enzyme lactate LdhL, while the D-(-) lactate is formed by the enzyme LdhD. Some bacteria have the lactate racemase (LR), which converts L-(+)-lactic acid in D-(-) lactic acid [29]. Lactic acid can be produced by bacteria, some yeasts and some filamentous fungi.

The filamentous fungi (*Rhizopus*) are still used for the production of lactic acid, because the process needs of bioreactors, with high aeration rate, they are obligatorily aerobic. Moreover, the efficiency of production processes using these micro-organisms is low due to the lower reaction rate caused by mass transfer limitation [30]. The lactic acid production is mainly carried out by lactic acid bacteria.

Bioabsorbable Polymers: Medicine and Surgical Applications

Polymers of lactic acid can also be used in medicine because they are bioabsorbable and can be used in tissue regeneration, sutures and fractures fixation [31]. Biostable and bioabsorbable biomaterials are used to manufacture implants for supporting, replacement, augmentation and guiding of growth of tissues. Bioabsorbable implants are a better choice for applications where only the temporary presence of the implant is needed. Because of bioabsorption of such implants, there is no need for a removal operation after healing of the tissue and the risks of implant related, long-term complications are eliminated or strongly reduced. Reinforcing of bioabsorbable materials is necessary in order to develop strong, safe and small implants for fixation of bone fractures and connective tissue damage. Self-reinforced bioabsorbable polymeric implants have been used so far extensively in the treatment of traumas of the musculoskeletal system [32].

The use of biodegradable materials in medicine is interesting because they can be used as an implant and will not need a second surgical event for removal. For example, a fractured bone, fixated with a rigid, non-biodegradable stainless steel implant, has a tendency for re-fracture upon removal of the implant. The bone does not carry sufficient load during the healing process, because the load is carried out by the rigid stainless steel [33]. However an implant prepared from a biodegradable polymer can be engineered to degrade at a rate that will slowly transfer load to the healing bone [34]. Another important application for biodegradable polymers is the possibility of using them as basis for drug delivery, alone, or in conjunction to functioning as a medical device. In orthopedic applications, the delivery of a bone morphogenic protein may be used to speed the healing process after a fracture [35], or the delivery of an antibiotic may help to prevent osteomyelitis following surgery [36]. The complications arising from biodegradable orthopedic implants of polymers of lactide typically occur at a rate of less than 10% [34].

Lactic Acid Bacteria

Lactic acid bacteria (LAB) are widely distributed in nature and occur naturally as indigenous microflora in a raw milk, drinking yoghurt and many others. They are Gram positive bacteria, non-motile, catalase negative, non-spore forming, cocci, coccobacilli or rods that have less than 55mol% G+C content in their DNA. They are organisms that ferment carbohydrates to form chiefly lactic acid [37].

For many centuries, LAB had been an effective form of natural preservation. In addition, they strongly determine the flavor, texture and frequently, the nutritional value of food and feed products [38,39]. The main genera of LAB include *Lactobacillus, Lactococcus, Leucconostoc, Pediococcus* and *Streptococcus*, as well *Aerococcus, Carnobacterium, Enteroococcus, Oenococcus, Teragenococcus, Vagococcus* and *Weissella* [40].

During the sugar fermentation, the bacteria can produce only L(+) lactic acid or only D(-) lactic acid and can as well produce equivalent concentration of both isomers. The most LAB produces both forms of lactic acid. The type of isomer that will be produced depends on the enzyme in the bacteria. If it has LdhL, it will produce L(+) lactic acid, or if it has LdhD it will produce D(-) lactic acid. Some bacteria have the racemase lactato, which converts L(+) lactic acid on D(-) lactic acid.

The LABs can be grouped according to the major isomer produced. This way, if they produce more than 75% of L(+) lactic acid are called L. Bacteria that produces more than 75% of D(-) lactic acid will be called D. And those that produce both isomers are called DL. The metabolism of carbohydrates by LAB can be called homofermentative or heterofermentative. In the first case, the sugar will be converted totally in lactic acid. In the case of heterofermentative the bacteria metabolism of glucose will produce lactic acid, ethanol and CO_2 (Figure 3) [41].

For identification of LAB two ways can be used: biochemical and molecular. When we use the biochemical way we consider the morphological, physiological and biochemical characteristics.

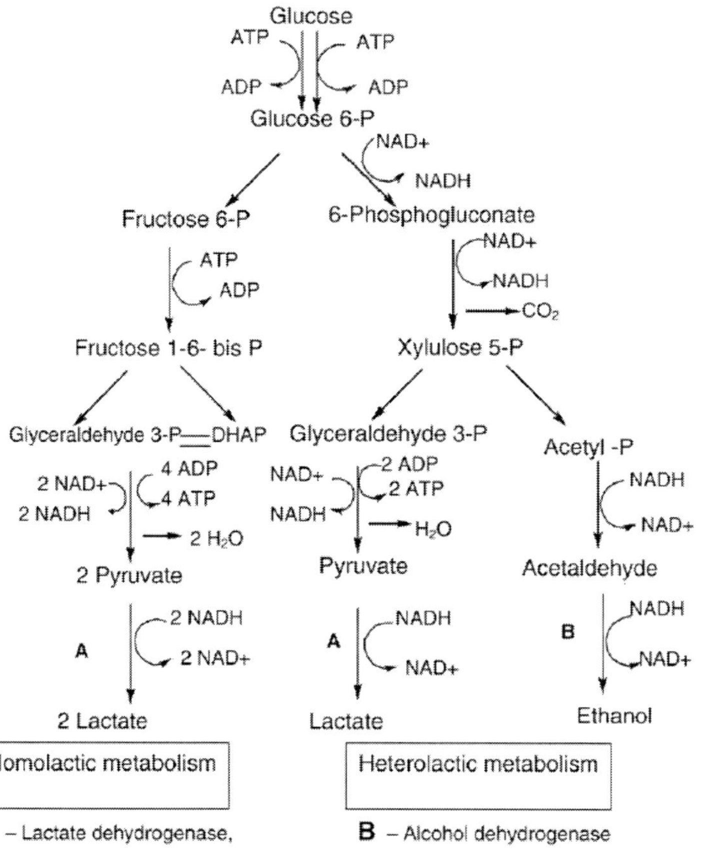

Figure 3. Metabolism of lactic acid bacteria.

The LAB are evaluated about the Gram, production of catalase, cytochrome-oxidase, hydrogenperoxidase, growth at 15 °C or 45 °C, at pH 4.4 or 9.6, production of lactic acid when many carbohydrates are used (L-arabinose, cellobiose, D-fructose, D-galactose, esculin, lactose, maltose, melizitose, melibiose, mannitol, D-mannose, raffinose, rhamnose, D-ribose, salicin, sorbitol, sucrose, trehalose and D-xylose), production of gas from 1% glucose, methyl red test and Voges-Proskauer, production of ammonia from arginine, nitrate reduction, indole production and growth on acetate agar [42].

The molecular way of identification considers the 16S rRNA sequences of LAB. When we use the two ways for the identification, the results are more reliable. *Streptococcus* were the generic name used to describe the bacteria that make chains in Gram coloration [43]. *Lactobacillus* are the heterogeneous group with rods positives Gram coloration, aerobic or aerotolerant, non-sporeforming, catalase negative. They depend on a complex nutritional, with vitamins, aminoacids for growth. They can grown in different habitats like, plants, soil, water, fermented foods (milk, beef, vegetables), cereals, beer, fruit, fish, in process of sugar. They are acidophilus, and can grown at pH 4,0 [43-45]. *Leuconostoc* are heterofermentative and produce D(-) lactic acid from glucose [46]. *Pediococcus* are coccus and homofermentative, and the major species produce DL lactic acid from glucose [43]. *Lactococcus* are positive gram, L(+) lactic acid producer from lactose, grown between 20 °C and 30 °C after 10 and 20 h of fermentation. They have complex nutritional requirements, are anaerobic, but can survive in presence of oxygen [47,48]. *Enterococcus* are facultative anaerobic, have homofermentative metabolism, and produce the isomer L(+) as a predominant isomer from glucose. They need a complex nutritional complex, with vitamins, biotin, and many aminoacids. They can grow at 10 °C and 45 °C, in medium with 6,5% NaCl and pH 9,6 [48,49]. The *Carnobacterium* produces L(+) lactic acid and it is similar to lactobacillus, but they do not grow in acetate. The *Tetragenococcus* needs NaCl to grow and produces L(+) lactic acid. The *Vagococcus* grows in aquatic habitat, and produces L(+) lactic acid from glucose [47].

Production of Lactic Acid from Renewable Resources

Biomass is considered an agro-renewable energy source. Around 3.5 million tonnes of agricultural waste are produced each year worldwide. The use of a particular carbohydrate raw material depends on its price, availability and purity. Although the agro-industrial residues are rich in carbon, its use is limited due to their low protein concentration [23].

The carbon source for microbial production of lactic acid can be sugar in its pure form, such as glucose, sucrose, lactose. However, these sugars are very expensive. Thus, different food products and agro-industrial waste or alternatives are cheaper than refined sugars to lactic acid production, such as whey [50], potato starch, tapioca, corn, wheat, cassava, beans, barley and carrot [51-53], molasses [54, 55] and sugarcane juice [56], cassava wastewater [57], corn steep liquor [54-57], office waste paper [58], hemicelluloses and lignocelluloses hydrolysates [59], cellulose [60], sweet sorghum, [61] and glycerol [62].

Table 1. Lactic acid production from different carbon sources

Organism	Substrate	LA (g L^{-1})	YLA/tot g/g	Ref
Lb. plantarum Lmism6	Sugarcane molasses	63.4	0.90	[54]
Lb. amylophilus ATCC 49845	hydr corn starch	33	0,73	[69]
Lactobacillus sm8	Whey	52.37	0.87	[50]
Lb. rhamnosus B103	Sugarcane juice	85.06	0.85	[56]
Lb. rhamnosus B103	Cassava wastewater	41.65	0.83	[12]
Lactobacillus lactis mutant RM2-24	Cellobiose	80	0,8	[70]
Lb. delbrueckii sp. *bulgaricus* NRRL B-548	Cellulose	52	0.58	[71]
R. oryzae GY18	Sucrose	97.5	0.81	[72]
R. oryzae GY18	Xylose	54.2	0.9	[72]
Lc. lactis sp. *lactis* ATCC 19435	hydr wheat flour, SSF	107	0.91	[73]
Lc. lactis IFO 12007	potato starch	25	0.5	[74]

Abbreviations: LA: lactic acid; YLA/tot: yield of g LA per g substrate provided; hydr: hydrolysed.

Sugarcane bagasse is used as a carbon source for lactic acid production by *R.oryzae* and *Lactobacillus* in solid state fermentation (SSF), hydrolyzed starch as a carbon source [63]. Generally, lactic acid bacteria (BL) are deficient in cellulolytic and amylolytic enzymes and require a prior acid or enzymatic hydrolysis of cellulose and starch [64].

Woiciechowski *et al.* [65] studied the economics of recovery of reducing sugars by acid and enzymatic hydrolysis of cassava bagasse. Although both, acid hydrolysis and enzymatic hydrolysis, have almost the same efficient based on the yield of reduction, acid hydrolysis is more economically advantageous because the energy consumption cost of enzymatic hydrolysis is 25% higher than the acid hydrolysis [65].

Another very significant waste of the dairy industry is cheese whey containing lactose as carbon source, protein, vitamins and minerals [50].

Molasses also have been studied as an abundant by-product, approximately 17.9 million tons of sugarcane molasses are produced per year in Brazil during the sugar manufacturing. In addition to its rich composition in fermentable sugars [66], nitrogen and vitamins, they have low cost and high availability in Brazilian territory. However, most results on the lactic acid produced (17- 90 g L^{-1}) and productivity (2.0 to 4.0 g L^{-1} h^{-1}) in batch fermentations is relatively low [67]. So, additional treatment of molasses with sulfuric acid, tri-calcium phosphate, potassium ferrocyanide and EDTA is necessary to increase efficiency in the fermentation, because the molasses contain heavy metals (iron, zinc, copper, manganese), and magnesium, calcium, which can inhibit cell growth, affect the pH and inactivate enzymes associated with the formation of the product [68].

FERMENTATION PROCESSES

Lactic acid production is usually produced in batches, but other processes are used, such as fed batch and continuous processes. According to Hofvendahl and Hahn-Hagerdal [75], the batch process leads to higher concentrations of product in most cases because there is an almost total consumption of the substrate, whereas in continuous processes remains a residual concentration. On the other hand, the continuous process usually leads to higher yields, especially if conducted at high dilution rates.

The best results obtained in the batch fermentation of lactic acid with glucose as the substrate are 150.2 g L^{-1} for the final L-lactic acid concentration and 1.34 g L^{-1} h^{-1} for the productivity, according to Bai *et al.* [76]. Yamane and Shimizu [77] defined as a fed batch culture, the addition of one or more nutrients. The fed batch process is successful; it removes the inhibition of the substrate, resulting in higher productivity, higher product concentration and improvement in the rate of cell growth [78,79].

There are several methods of feeding, the flow can be constant or variable, pulse fed batch, fed batch activated by constant residual glucose concentration and exponentially fed batch [80]. Bai *et al.* [76] developed a process for production of ammonium lactate by *Lactobacillus lactis* BMES M-18 in fed batch fermentation at controlled pH. They observed that the lactate production of ammonia was markedly improved by continuous feeding of glucose and concentration product, the maximum dry weight of cells and the average productivity obtained were 161.2, 2.15 g L^{-1} and 2.02 g. L^{-1} h^{-1} respectively.

Ding and Tianwei [81] studied various methods of lactic acid fermentation by *L.casei* LA-04-1 in order to increase the yield of L(+) lactic acid. A solution of glucose (850 g L^{-1}) and yeast extract (1%) were pumped into the fed batch fermentation in the exponential method. The maximum production of lactic acid (210 g L^{-1}) and L-(+) lactic acid (180 g L^{-1}) were obtained in exponentially fed batch with glucose and yeast extract, the maximum cell concentration (4.30 g L^{-1}) was obtained after 70.5 h of fermentation, the yield and productivity of L-(+) lactic acid was 90.3% and 2.14 g L^{-1} h^{-1}, respectively.

Several reasons for the utilization of fed batch processes are cited in the literature [77,82], such as a decrease of catabolic repression in the rapidly metabolizable carbon sources repress the expression of genes encoding enzymes related to metabolism, prevention of inhibition by substrates, in which any nutrient source can become inhibitory, depending on the concentration of the medium, the micro-organism and fermentation conditions, minimizing the formation of toxic products of metabolism.

Fedbatch culture proved to be a better fermentation system for the production of lactic acid than batch culture in a system with mixed cultures of free and co-immobilized *Lcasei* and *L.lactis* [83]. The maximum lactic acid concentration (46 g L^{-1}) in fedbatch culture was obtained with both free cell mixture and co-immobilized cells at a substrate concentration of 100 g L^{-1} and a feeding rate of 250 mL h^{-1}. Another way to achieve high productivity is by recycling the cells, resulting in a high cell mass without reducing the yield. Immobilization of cells usually, to increase the cell density getting the better productivity and yield, the co-immobilized cells in Ca-alginate beads retained their ability to produce lactic acid for 20 days. On the other hand, recirculation of cells gave higher lactic acid concentrations and higher or equal yields and in several studies, better results were obtained by using free cells [75].

Other type of fermentation processes is the solid-state fermentation (SSF), defined as the growth of microorganisms on moist solid materials in the absence of free-flowing water [84,85]. Several works have been peformed on the production of industrial enzymes using SSF. In SSF, the solid substrate not only supplies nutrients to the culture but also serves as an anchorage to the microbial cells. A study was made to develop a novel technology for L(+) lactic acid production by SSF using *L. amylophilus* GV6 culture for which wheat bran (a by-product of wheat milling industry) was selected as solid substrate and support [19,86].

CURRENT TRENDS AND FUTURE PROSPECTS IN RESEARCH ON LACTIC ACID

The world market for lactic acid, virtually immune to the recent turmoil in the global economy is currently experiencing steady growth. It was noted an increase in demand due to the emergence of new applications. It is expected to continue growth in the next years. According to a new report by Global Industry Analysts Inc, the global market for lactic acid will reach 328,900 tonnes in 2015.

Lactic acid is used in foods and beverages, cosmetics, pharmaceuticals, biodegradable plastics, and other chemical sectors. The use of lactic acid in personal care and pharmaceuticals, such as intravenous solutions, moisturizers, shampoos, creams and anti aging skin soaps are still high in many countries.

It is expected that demand for the use of lactic acid increase further due to the global call for the use of biopolymers in detriment of conventional polymers with respect to the environment, and also because of easier recycling and low cost.

Thus, we can highlight the polylactic acid (PLA) biodegradable polymer as a key. Another area of focus, responsible for increasing demand for lactic acid production would be green solvents, are esters of low molecular weight lactic acid derivatives such as ethyl, propyl, butyl and lactate. The Environmental Protection Agency (EPA) approved the lactate ester solvents as inert ingredients with negligible toxicity, and as an excellent environmental profile [87].

Moreover, lactic acid can potentially be used for the manufacture of high chemicals volume such as propylene glycol, propylene oxide, acrylic acid, acrylic esters and other chemical intermediates [87]. With the increase on lactic acid demand and on environmental impact cautions of the gypsum accumulation as a by-product of traditional fermentative production of lactic acid, increased efforts to develop alternative technologies are being made. Genetic engineering approaches have been explored for the improvement of lactic acid bacteria (BL) [88].

Due to the intrinsic problems of growth under anaerobic conditions and inability of BL to produce lactic acid under low pH, the big challenge is in developing BL with tolerance to low pH and inhibition by the accumulation of the product, making recovery of lactic acid easier and resulting in less accumulation of gypsum.

The use of lactic acid to produce polilactates for employment in health has been quick with different applications, both as suture, artificial prosthesis to replace the titanium, and lately, as a stent in place of metals. Despite considerable pioneering effort invested by some companies in developing and marketing products based on lactic acid production technology of lactic acid, more studies in advanced and implemented technology are necessary.

References

[1] Ohara, H. *Applied Microbiology and Biotechnology*, 62: 474–477 (2003).

[2] Di Lorenzo, M. L. *European Polymer Journal*, 41: 569–575 (2005).

[3] Ryu, H. W., Yun, J. S., Wee, Y. J. (2003), In A. Pandey (ed.), Concise encyclopedia of bioresource technology, lactic acid (p. 635). New York: The Haworth Press.

[4] Lipinsky, E. S., Sinclair, L. G. *Chemical Engineering Progress*, 82: 26–32 (1986).

[5] Hofvendahl, K., Hahn-Hägerdal, B. *Enzyme and Microbial Technology*, 20: 301–307 (1997).

[6] Ikada, Y., Jamshidi, K., Tsuji, H., and Hyon, S. H. *Macromolecules*, 20: 904–906 (1987).

[7] Tsuji, H., Hyon, S. H., Ikada, Y. *Macromolecules*, 24: 5651–5656 (1991).

[8] Fitzpatrick, J. J., Keeffe, U. O. *Process Biochemistry*, 37: 183–186 (2001).

[9] Tejayadi, S., Cheryan, M. *Applied Microbiology and Biotechnology*, 43: 242–248 (1995).

[10] Calabia, B. P., Tokiwa, Y. *Biotechnological Letters*, 29: 1329–1332 (2007).

[11] Dumbrepatil, A., Adsul, M., Chaudhari, S., Khire, J., Gokhale, D. *Applied and Environmental Microbiology*, 74: 333–335 (2008).

[12] Coelho, L. F., De Lima, C. J. B., Bernardo, M. P., Alvarez, G. M., and Contiero, J. *Journal of the Science of Food and Agriculture*, 90: 1944–1950 (2010).

[13] Buyukkileci, A.O., Harsa, S. *Journal of Chemical Technology and Biotechnology*, 79: 1036–1040 (2004).

[14] Yu, L., Lei, T., Ren, X., Pei, X., Feng, X. *Biochemical Engineering Journal*, 39: 496–502 (2008).

[15] De Lima, C.J.B., Coelho, L.F., Blanco, K.C., Contiero, J. *African Journal of Biotechnology*, 8: 5842–5846 (2009).

[16] Narayanan, N., Roychoudhury, P.K. and Srivastava, A. *Eletronic Journal of Biotechnology*, 7: 167-179 (2004).

[17] Hofvenahl, K., Hahn-Hagerdal, B. Enzyme and Microbial Technology, 26: 87-107 (2000).

[18] Tanaka, T., Hoshina, M., Tanabe, S., Sakai, K., Ohtsubo, S.,Taniguchi, M. *Bioresource Technology*, 97: 211-217 (2006).

[19] Naveena, B.J., Altaf, Md., Bhadriah, K., Reddy, G. *Bioresourse Technology*, 96: 485-490, (2005).

[20] Auras, R.A., Harte, B. and Selke, S. Poly (lactide): moisture sorption characteristics and storage consequences. In: ANTEC 2003 THE 61 ST *Annual Technical Conference of the Society of Plastics Engineers*, 61: 4-8, (2003).

[21] WIKIPÉDIA. O ácido polilático. Wikimedia Foundation. Disponível em: <http://en.wikipedia.org/w/index.php?title=Polylactic_acidandoldid =407314317>. Acesso em: 19 fev. 2011.

[22] Sodegard, A. and Stolt, E M. *Progress in Polymer Science*, 27: 1123–1163, (2002).

[23] Pondey, A.; Soccol, C. R.; Rodriguez-Leon, J. A.; Nigan, P. Solid state fermentation in biotechnology: fundamentals and applications. *Asiatech Publishers*, New Delhi, 2001.

[24] Zhou, S., Shanmugam, K.T.; Yomano, L.P.; Grabar, T. B., Ingram, L.O. *Biotechnology Letters*, v. 28, 663–670, 2006.

[25] Buchta, K. Lactic acid. In: H.J. Rhem and G. Reed, Editors, Biotechnology. Volume 3. Biomass, *Microorganisms for Special Applications, Microbial Products I, Energy from Renewable Resources*, V.C.H., Weinheim (1983), pp. 409–417.

[26] Garvie, E. I. *Microbiology Reviews*, 44: 106-139, (1980).

[27] Kandler, O. and Weiss, N. Lactobacillus. In: Sneath, P. H. A.; Mair, N. S.; Sharpe, M. E.; Holt, J. G. *Bergey's manual of systematic bacteriology*. Baltimore: Williams and Wilkins, 1986. p. 1209-1234.

[28] Schleifer, K.H. (1986) Gram positive cocci. I*n Bergey's Manual of Systematic Bacteriology*, vol. 2. ed. Sneath, P.H.A., Mair, N.S. and Holt, J.G. pp. 1043–1074. Baltimore: Williams and Wilkins Press.

[29] Kandler, O. Antonie van Leeuwenhoek, Amsterdam, 49: 209-224, (1983).

[30] Park, E. Y., Kosakai, Y. and Okabe, M. *Biotechnology Progress*, 14: 699–704, (1998).

[31] Sakata, M.M., Rincon, M.C.A. do and Duek, E.A.R. Estudos da interação polímero/cartilagem/osso utilizando Poli (ácido lático – co – ácido glicólico) e Poli (p-Dioxanona) em condilo femural de coelhos. Polímeros: Ciência e Tecnologia, 24: 176-180 (2004).

[32] Örmälä, P. T.; Pohjonen, T.; Rokkanen, P. Bioabsorbable polymers: Materials technology and surgical applications. In: Proceedings of the Institution of Mechanical Engineers, Part H: *Journal of Engineering in Medicine* February 1, 212: 101-111, (1998).

[33] Middleton, J.C; Tipton, A.J. *Biomaterials*, 21: 2335-2346, (2000).

[34] Athanasiou KA, Agrawal CE, Barber FA, Burkhart SS. Arthrosc: *J. Arthrosc. Relat. Surg.*; 14: 726-37, (1998).

[35] Wang EA, Rosen V, D'jAlessandro JS, Bauduy M, Cordes P, Harada T, Isreal DI, Hewick RM, Kerns KM, LaPan P, Luxenberg D, McQuaid D, Moutsatsos IK, Nove J, Wozney JM. *Proc. Natl. Acad. Sci.*, 87: 2220-4, (1990).

[36] Ramchandani M, Robinson D. *J. Controlled Rel.;* 54:167-75, (1998).

[37] Schleifer, K.H., Ludwig. W. Phylogeny of the genus LLX~O~Land/~~S and related genera. *System. Appl. Microbial.* 18: 461 467, (1995).

[38] Lee, B. bacteria-Based Process and products. In: *Fundamentals of Food Biotechnology*, Lee, B.(Ed), Wiley-Inter Science, New York, pp 219-290.

[39] Tserovska, L., Stefanova, S.; Yordanova. Identification os lactic acid bactéria isolated from Katyk, goats Milk, and Cheese. *J. Cult. Collect.*, 3: 48-52, (2002).

[40] Axelsson L. Lactic acid bacteria: classification and physiology. In: Salminen S, von Wright A, Ouwehand A, editors. Lactic acid bacteria: *microbiological and functional aspects*. 3rd rev. and exp. ed.New York: Marcel Dekker, Inc.; 2004. p. 1-66.

[41] Lehninger, A.L.; Nelson, D.L.; Cox, M.M. *Principios de Bioquimica*, 3ª ed. Sarvier, 2002.

[42] Nair, P.S, Surendran, P. K. *Journal of Culture Collections*, 4: 48-52, (2005).

[43] Stiles, M.E.; Holzapfel, W.H. *International Journal of Food Microbiology*, 36: 1-29, (1997).

[44] Huang, L.P.; Jin, B.; Lant, P.; Zhou, J. *Journal of Chemical Technology and Biotechnology*, 78: 899–906, (2003).

[45] Sneath, P.H.A.; J.G. Holt. *Bergey's Manual of Systematic Bacteriology*, 2nd edition. Vol. 1, p. 64. A Waverly Company, Williams eWilkins, Springer-Verlag, NY, USA, 2001.

[46] Daeschel, M.A.; Anderson, R.E.; Fleming, H.P. *FEMS Microbiology Reviews*, 46: 357-367, (1987).

[47] Holt, J.G.; Krieg, N.R.; Sneath, P.H.A.; Staley, J.T.; Williams, S.T. (1994) *Bergey's manual of determinative bacteriology*, 9th edn. William and Wilkins, MD, pp 527- 564.

[48] Wood, B.J.B., Holzapfel, W.H. *The genera of lactic acid bacteria* (1st ed.).Glasgow: Blackie Academic and Professional.1995.

[49] Schillinger, U., Lucke, F.K. *Food Microbiology*, 4: 199-208, (1987).

[50] De Lima, C.J.B.; Contiero, J.; Coelho, L.F. *Food Technology and Biotechnology*, 48: 171-181, (2010).

[51] Rojan PJ, Nampoothiri KM, Nair AS, Pandey A. *Biotechnol Lett* 27:1685–1688, (2005).

[52] Narita, J., Nakahara, S., Fukuda, H. and Kondo, A. *Journal of Bioscience and Bioengineering*, 97: 423-425, (2004).

[53] Oh, H., Wee,Y.J, Yun, J.S., Han,S.H, Jung, S. and Ryu, H.W. *Bioresource Technology*, 96, 1492-1498, (2005).

[54] Coelho, L. F., Lima, C. J. B., Bernardo, M. P. and Contiero, J. *Brazilian Journal of Chemical Engineering*, 28: 27-36, (2011).

[55] De Lima, C.J.B.; Coelho, L.F.; Blanco, K.C.; Contiero, J. *African Journal of Biotechnology*, 8: 5842-5846, (2009).

[56] De Lima, C.J.B.; Coelho, L.F.; Silva, G.P.; Alvarez, G.M.; Contiero, J. *Journal of Microbial and Biochemical Technology*, 2: 64-69, (2010).

[57] Coelho, L.F., De Lima, C.J.B., Bernardo, M.P., Alvarez, G.M. and Contiero, J. *Journal of the Science of Food and Agriculture*, 90: 1944-1950, (2010).

[58] Park, E.Y, Anh, P.N. and Okuda, N. *Bioresource Technology*, 93: 77–83, (2004).

[59] Karel, M., Jaroslav, V., Vera, H., Mojmir, R. *J. Biotechnol.* 56:25–31, (1997).

[60] Venkatesh, K.V. *Bioresour. Technol.* 62:91–98, 1997.

[61] Richter K, Trager A. Acta Biotechnol; 14:367–78, (1994).

[62] Mazumdar, S., Clomburg, J.M. and Gonzalez, R. *Applied and Environmental Microbiology*,76: 4327-4336, (2010).

[63] Yu RU, Hang YD. *Biotechnol. Lett.* 11:597–600, (1989).

[64] Pandey A, Soccol CR. J Sci Ind Res 59:12–22, (2000).

[65] Woiciechowski AL, Nitsche S, Pandey A, Socool CR. *Braz. Arch. Biol. Technol.* 45: 393–400, (2002).

[66] Lima, U.A., Aquarone, E., Borzani, W. *Biotecnologia: tecnologia das fermentações*. São Paulo: Edgard Blücher, 1975. v.1.

[67] Kotzamanidis, C., Roukas, T., Skaracis, G. *World Journal of Microbiology and Biotechnology*, 18: 441–448, (2002).

[68] Roukas T. (1998). *Process Biochemistry*, v.33, p.805–810.

[69] Mercier P, Yerushalmi L, Rouleau D, Dochain D. *J. Chem. Technol. Biotechnol.* 55:111–21 (1992).

[70] Singhvi, M., Joshi, D., Adsul, M., Varma, A., Gokhale, D. *Green Chem.*, 12: 1106-1109 (2010).

[71] Venkatesh KV, Wankat PC, Okos MR. AIChE *Symp. Ser.*; 300: 80 –7 (1994).

[72] Guo, Y.; Yan, Q.; Jiang, Z.; Teng, C.; Wang, X. *J. Ind. Microbiol. Biotechnol.* 37:1137–1143, (2010).

[73] Hofvendahl K, Hahn–Hagerdal B. *Enzyme Microb. Technol.* 20:301–7 (1997).

[74] Hofvendahl, K.; Hahn-Hagerdal, B. *Enzyme and Microbial Technology, Philadelphia*, v. 26, p. 87-107, 2000.

[75] Bai DM, Wei Q, Yan ZH, Wei Q, Zhao XM, Li XG, Xu SM. *Biochem. Eng. J.*, 19:47–51, 2004.

[76] Yamane, T.; Shimizu, S. *Advances in Biochemical Engineering Biotechnology*, 30: 147-194, 1984.

[77] Hujanen, M., Linko, S., Linko Y.Y, Leisola M. *Applied microbiology and biotechnology*. 56: 126-130, 2001.

[78] Korz D.J., Rinas, U.; Hellmuth, K., Sanders, E.A., Deckwer W.D. *Journal of Biotechnology*, 39: 59-65, 1995.

[79] Keller, R. Dunn, I J. *Journal of Chemical Technology and Biotechnology*, 28: 784-790, 1978.

[80] Ding, S.; *Process Biochemistry*, 41: 1451-1454, 2006.

[81] Schmidell, W.; Lima, A. U.; Aquarone, E.; Borzani, W. *Biotecnologia Industrial.* São Paulo: E. Blücher, 2001. v. 2, 254 p.

[82] Roukas, T. and Kotzekidou, P. *Enzyme and Microbial Technology*, 13: 33-38 (1991).

[83] Moo-Young, M., Moreira, A.R. and Tengerdy, R.E. 4: 117-144 (1983).

[84] Pandey, A. *Biochemistry*, 27: 109-117 (1992).

[85] Altaf, M.; Naveena, B.J.; Venkateshwar, M.; Kumar, E.V.; Reddy, G. *Process Biochemistry*, 41: 465-472, (2006).

[86] Datta R, Henry M. *J. Chem. Technol. Biotechnol.* 81: 1119–1129, (2006).

[87] Singh SK, Ahmed SU, Pandey A. *Proc. Biochem.* 41:991–1000, (2006).

In: Lactic Acid
Editors: A. Migallón and R. Ruseckaite
ISBN: 978-1-62081-643-1

Chapter 3

SILOXANE-CONTAINING VATERITE / POLY (LACTIC ACID) FIBREMATS WITH IMPROVED DUCTILITY

Akiko Obata and Toshihiro Kasuga*
Department of Frontier Materials, Graduate School of Engineering, Nagoya Institute of Technology
Gokiso-cho, Showa-ku, Nagoya, Japan

ABSTRACT

Siloxane-containing vaterite (SiV)/poly (L-lactic acid) (PLLA) hybrid fibremats (SiPVHs) have outstanding potential for bone regeneration applications. SiPVHs were prepared by using melt-blending and electrospinning protocols and modified by increasing PLLA content from 40 to 70 wt% (denoted by SiPVH60 and SiPVH30, respectively). The tensile strength of SiPVH30 was 10 times higher than that of SiPVH60. The elongation before the failure of SiPVH30 was twice as large as that for SiPVH60. The necking of fibers occurred at the failure points in SiPVH30 but not in SiPVH60. Ionic silicon-species in Tris-buffer solution from SiPVH30 was the relatively constant release. SiPVH30 could also be coated with hydroxyapatite by being soaked in simulated body fluid. A new type of membrane was prepared by bonding SiPVH60 fibremat with PLLA. The bilayered membrane showed

* kasuga.toshihiro@nitech.ac.jp.

flexibility, relatively high mechanical strength to allow easy handling, and not only functions as the barrier to control the intrusion of soft tissue but also the function of a 3-D scaffold which enhances the cell activation and the bone formation.

Keywords: Siloxane, Calcium, Poly(lactic acid), Fibremat, Electrospinning

INTRODUCTION

There are many materials for bone reconstruction on the market. Calcium phosphate ceramics such as hydroxyapatite and tricalcium phosphate and silica-based glasses such as Novabone® (bioactive glasses) are very popular[1]. They have been used clinically as bone graft substitute materials as they can bond to natural bone directly. Composite materials out of such ceramic or glass and a biodegradable polymer have been widely investigated to create materials with bioactivity, high toughness and flexibility[2-4]. Biodegradable polymers used clinically are poly (L-lactic acid) (PLLA) and poly(glycolic acid) (PGA), etc. Calcium phosphate ceramics or bioactive glasses can release calcium and phosphate ions, which contribute to the formation of hydroxycarbonate apatite (HCA) and new bone, through the biodegradation of the polymer in living body.

Scaffolds out of the biodegradable polymer have been developed for use in bone reconstruction, tissue engineering and drug delivery systems[5,6]. Three-dimensional and porous scaffolds are common and regarded to be useful in biomaterials. The pores in the scaffolds are regarded to interconnect with each other and have pore diameters of at least 100 μm to allow cell migration, bone ingrowth and eventually vascularization [7,8]. Various types of porous scaffolds out of calcium phosphate ceramics, bioactive glasses and their composites with biodegradable polymers have been developed.

Electrospinnig is one of the processes that can generate polymer fibremats with high porosity and flexibility, both essential components of the scaffolds[9-11]. The process of electrospinning is affected by both system parameters and process parameters. System parameters include polymer molecular weight and polymer solution properties, such as viscosity and conductivity. Process parameters, on the other hand, involve the flow rate of the polymer solution, electric potential and the distance between the capillary and collector, among others [9]. Fibre diameter and thickness of fibre mesh scaffolds can be controlled by optimizing these parameters.

It is necessary to choose the appropriate materials for optimizing the mechanical and biomimetic properties of a fibre mesh, since the orientation and diameter of the fibres in electrospun scaffolds can influence cellular compatibility [12-14]. The fibre diameters in electrospun scaffolds have, for example, been reported to influence osteoblast proliferation and mineralization.

Guided bone regeneration (GBR) is one of the techniques for regenerating bones in particular to achieve the dental implant. When a dental implant is fixed in jaw bone, if thickness of alveolar bone is not enough for achieving the implantation, bone regeneration is needed. A defect formed in alveolar bone for the implantation is covered with a membrane, so called "GBR membrane", is prone to prevent the intrusion of soft tissue such as gingival epithelial and connective tissues into the defect and to keep enough space for the proliferation and mineralization of osteogenic cells. The bone regeneration in the defect should be accelerated by the selective growth of the osteogenic cells there. However, it usually takes for about six months to completely reconstruct the alveolar bone, even if the GBR is applied. The GBR membranes require flexibility to adapt to the bone shape with defect and strength to maintain the space for bone formation and to connect with the soft tissues. They require an additional property, enhancement of the bone formation.

Ionic dissolution products from inorganic biomaterials have been attracting attention and regarded as a key to understand bioactivities of the materials in body. Many elements, calcium, phosphorus, strontium, copper, zinc, boron, vanadium, magnesium and cobalt, have been reported to have stimulatory effects on osteogenic cells[15]. Silicon species is one of the essential, trace elements for healthy bone and have been recently reported to stimulate functions of osteogenic cells[7, 16-20]. This has been supported by the reports about bioactive glasses, such as Bioglass® (45 % SiO_2, 24.5 % CaO, 24.5 % Na_2O, 6 % P_2O_5). Ionic dissolution products released from Bioglass® stimulated osteogenic cells at the genetic level. Xynos *et al.* reported that some gene expressions in human osteoblasts were stimulated by silicon species in a culture medium, derived from Bioglass®[17,18]. They also reported that the expression of insulin-like growth factor II in human osteoblasts was enhanced by silicon species in the culture medium, which stimulated the cells to proliferate. Reffitt*et al.* reported that orthosilicic acid in a culture medium was effective to stimulate collagen type I synthesis and osteoblastic differentiation in human osteoblast-like cells [20].

Our previous studies reported that a composite material of PLLA and siloxane-containing calcium carbonate (vaterite) (SiV) exhibits outstanding potentials for bone regeneration applications as it enhances the proliferation and mineralization of osteoblast-like cells *in vitro* and the formation of mineralized tissue *in vivo*[21, 22]. This composite material releases a trace amount of silicon in a cell culture medium and shows an HCA-forming ability in simulated body fluid in one day. Osteoblast-like cells (MC3T3-E1 cells) proliferated and mineralized on the composite membrane.

Fibremats out of the composite material were successfully prepared by an electrospinning method [22-25]. SiV should play an important role in the enhancement of their osteoconductivity, since it releases both calcium and silicon ionic species. We have found that it is possible to prepare the composite material containing ~60 wt% of SiV in the system without the aggregation of SiV particles in a polymer matrix. Too much SiV, however, should cause a brittleness of the composite material. In spite of the success of the fibremat, mechanical properties such as tensile strength and ductility (in the present work, ductility refers to the elongation of the fibremats before failure) are still required to be improved for the applications. The present communication shows the fibremat out of the SiV/PLLA composite material improved in mechanical properties by manipulating its composition or incorporating with a PLLA membrane.

SILOXANE-CONTAINING VATERITE

Silicon-containing vaterite (SiV) powders were prepared by a carbonation process with methanol. A sample of 150 g of $Ca(OH)_2$ was mixed with 60 mL of aminopropyltriethoxysilane (APTES) and 2000 mL of methanol with CO_2 atmosphere for 75 min at 2000 mL min^{-1}. The resulting slurry was dried at 110 °C, resulting in the SiV powders.The amount of silicon in SiV was estimated to be approximately 2 wt% by X-ray fluorescence analysis. Results of X-ray diffractometry indicated that SiV consisted of only vaterite crystalline phase. The specific surface area of SiV was estimated to be 67.2 m^2g^{-1} according to BET theory using nitrogen gas adsorption.

The morphology of SiV particles was observed by scanning electron microscopy (SEM) and transmission electron microscopy (TEM). SiV particles were aggregates consisting of several primary particles with the size of several hundreds nm (Figure 1). The TEM images in Figure 2 show the morphologies of the SiV particles. The low-magnification TEM image (Figure 2(a)) showed the particle size of the SiV, which is approximately 1 μm and these results were consistent with the literature[21]. The high-magnification

TEM image (Figure 2(b)) revealed the nanostructure of the SiV particles, consisting of smaller primary nanoparticles with diameters approximately 20 nm.

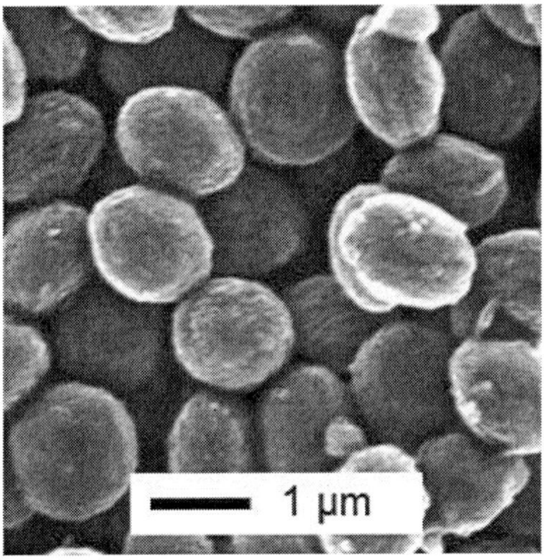

Figure 1. SEM image of SiV particles.

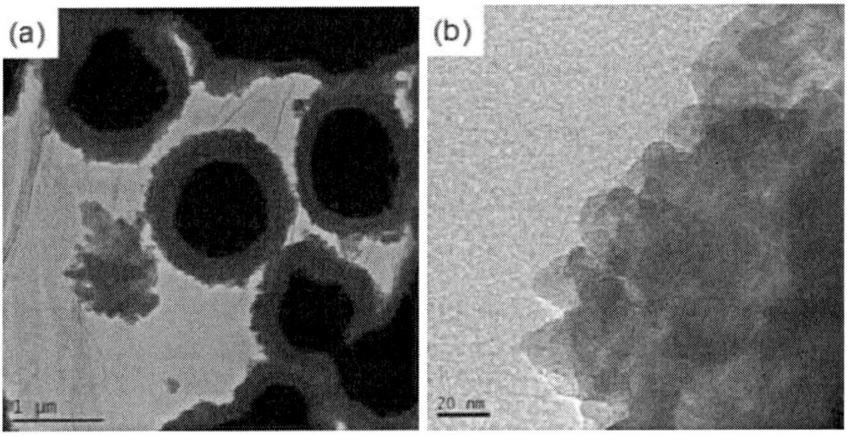

Figure 2. TEM images of SiV particles. (a) lower magnification, (b) higher magnification.

SiV/PLLA Composite Fibremats

Electrospinning of Composite Materials

The fibremats were prepared by melt-blending 40 wt% PLLA (molecular weight (M_W): 160 kDa) with 60 wt% SiV particles at 200 °C for 10 min. The products after the melt-blending were dissolved in chloroform to prepare 10 wt% solution, which was used to conduct electrospinning at 20 kV to obtain siloxane/poly(lactic acid)/vaterite hybrid fibremats (denoted by SiPVHs). In order to modify SiPVHs, the SiV particles content was changed from 60 wt% to 30 wt% (denoted by SiPVH60 and SiPVH30, respectively). The morphology of samples was observed by SEM. Comparison between the two materials was conducted in terms of mechanical properties and ion releasability.

SEM images in Figure 3 show the morphologies of SiPVH30 and SiPVH60. Microscopically, there were many pores (< 1 μm) in SiPVH30. Since chloroform is a relatively poor solvent for PLLA, the interaction between PLLA and chloroform is not sufficient to prevent phase separation in the SiPVH30 solution. The quick evaporation of pure chloroform phases could leave the pores shown in Figure 3(a). There were few pores in SiPVH60, which is due to the relatively lower polymer content in SiPVH60 and less evaporated chloroform during the electrospinning process. The primary fibre diameters of all the samples were found to be approximately 10 μm.

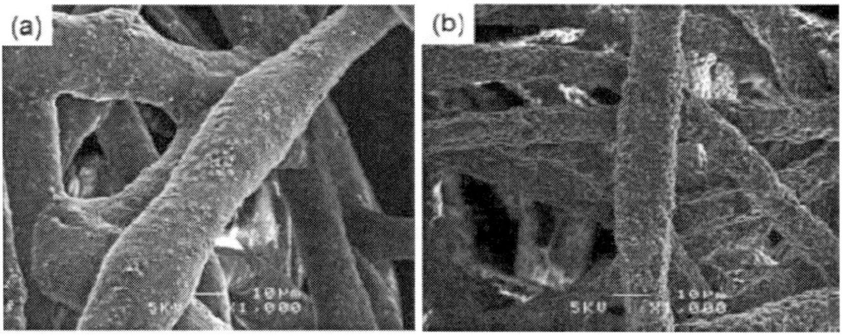

Figure 3. SEM images of (a) SiPVH30 and (b) SiPVH60.

The viscosity of chloroform solutions before the electrospinning process was measured with a viscosimeter. The differences in viscosity could also be explained by the different affinities between the chloroform and the polymer. The viscosity of SiPVH30 chloroform solution was 4582 Pa·s. The viscosity of the solution could affect the stabilisation of the Taylor cone during the electrospinning process. The PLLA chloroform solution was relatively less well-detangled and the entangled PLLA polymer chains increased the solution viscosity and enhanced the single-fibre electrospinning. This indicates that the affinity between the organic solvent and the polymer during the electrospinning process is critical to control the solution viscosity for electrospinning and in turn control the final fibre diameter.

Ion Release from Fibremats

The amount of silicon released from SiPVH30 or SiPVH60 fibremat in a buffer solution was measured. The fibremats of 15 mm in diameter with 100 ~ 120 μm in thickness were soaked in 4 mL of Tris buffer solution (TBS) at 37 ^{o}C for 5 days. The TBS media after 1, 3, and 5 days of soaking were refreshed and analysed by inductively coupled plasma atomic emission spectroscopy (ICP-AES). The silicon ion concentrations released from the fibremat in TBS are shown in Figure 4. Almost no ions released from SiPVH60 during the 4th and 5th days of the soaking test. SiPVH30 constantly released the ions. The ion release behaviour of the fibremat may be influenced by the composite structure in the fibres, that is, the condition of SiV particles in the polymer matrices. In SiPVH60, much SiV particles were exposed to TBS and immediately released the ions in 3 days after soaking. The particles might be in contact with each other in the fibre and there should be a lot of path for TBS to permeate into the inside of fibre. The amount of SiV particles exposed to TBS and such path in the fibre should be little in SiPVH30. The ions were expected to be released slowly from each sample with the polymer degradation when they were soaked in TBS for longer, since the amount of the ions released from each sample over 5 days of soaking was just a little bit of the total amount of the ions contained in the sample.

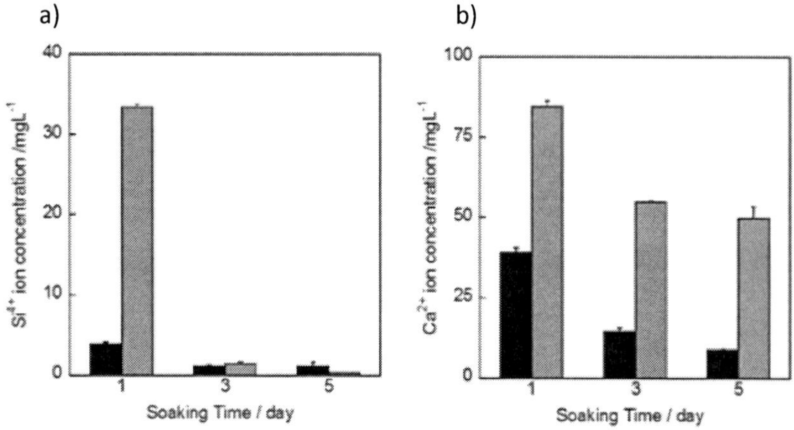

Figure 4. Concentrations of (a) Si4+ and (b) Ca2+ ions released from SiPVH30 and SiPVH60 after soaking in TBS for 5 days.

Mechanical Properties of Fibremats

In order to obtain fundamental data on mechanical properties, Japanese industrial standard tensile tests JIS L1015 were conducted in air at room temperature to measure their tensile strengths and elongation before failure (test details: span length, 20 mm; sample dimension, 40×5×0.1 mm^3; crosshead speed, 20 mm min^{-1}). The failure points of the samples after the tensile tests were observed by using SEM.

The tensile strength and elongation before failure of fibremats SiPVH30 and SiPVH60 are shown in Figure 5. The mechanical test results showed that the tensile strength and elongation before failure of SiPVH30 were 3.3 MPa and 3.0%, respectively, which were significantly higher than those of SiPVH60 (0.3 MPa and 1.6%).The SEM images in Figure 6 show the failure points of fibremats SiPVH30 and SiPVH60 after the tensile tests. Necking occurred at the failure points of SiPVH30, whereas no necking occurred at the failure points of SiPVH60 fibres.

Since PLLA is much more ductile than SiV particles, the PLLA matrix in the SiPVH30 fibres is suggested to induce the necking, which resulted in strain hardening and elongation of the fibres. The strain hardening and elongation of the fibres in SiPVH30 can in turn contribute to the enhanced tensile strength and enlarge elongation before failure, respectively. This indicates the high polymer content in the fibremats could enhance the mechanical properties of bone filler materials.

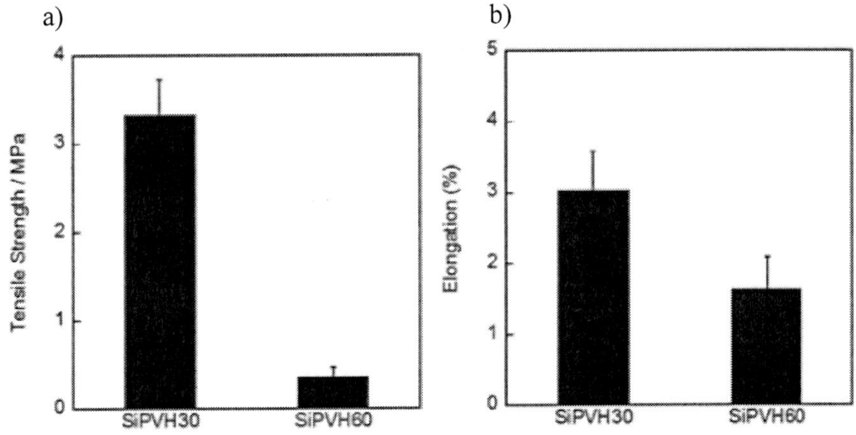

Figure 5.Mechanical properties of SiPVH30 and SiPVH60.(a) Tensile strength and (b) elongation before failure.

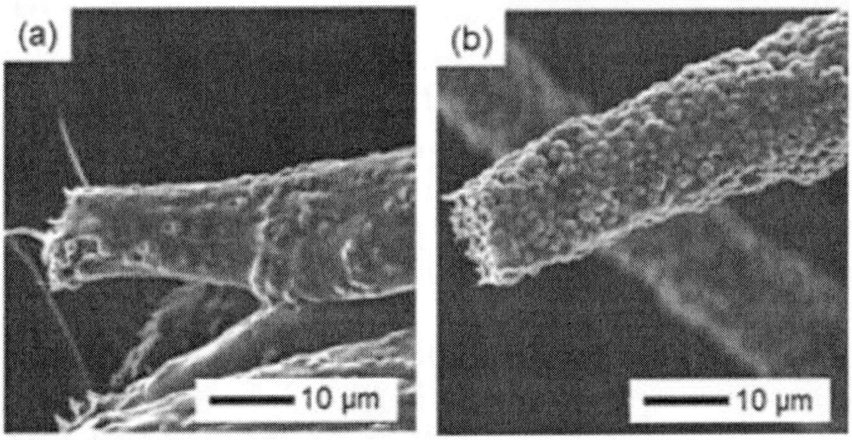

Figure 6. SEM images of the failure points of (a) SiPVH30 and (b) SiPVH60 after the tensile tests.

HCA-Coating on Fibre Surfaces

The fibremats coated with HCA can enhance cellular activities such as attachment and proliferation. HCA-forming on SiPVH30 and SiPVH60 fibremat was done by soaking in concentrated simulated body fluid (1.5 SBF) at 37 °C for 1 and 3 days. The 1.5 SBF consists of 3.75 mM of Ca^{2+}, 213.0 mM of Na^{+}, 2.25 mM of Mg^{2+}, 7.5 mM of K^{+}, 223.2 mM of Cl^{-}, 6.3 mM of HCO_3^{-}, 1.5 mM of HPO_4^{2+} and 0.75 mM of SO_4^{2-}, adjusted at pH 7.4 by including $(CH_2OH)_3CNH_2$ and HCl[26]. The surfaces of the fibremats after soaking were observed by SEM (Figure 7) to be almost completely covered with HCA, which was confirmed by XRD. Cauliflower-like deposits were partially observed on the fibre surfaces in SiPVH30 after 1day-soaking although the fibre surfaces in SiPVH60 were covered with the deposits. The fibre surfaces in SiPVH30 were completely covered with the deposits after 3days soaking. The HCA deposition obviously related to the content of SiV particles in the fibremat. In our preliminary experiment, no HCA deposited on the fibre surfaces in SiPVHfibremats with ~20 wt% of SiV particles, when they were soaked in 1.5SBF for 1day. We could conclude the SiPVH containing more than 30 wt% of SiV particles should be one of attractive candidates for biomaterials with enhancement of bone formation in body.

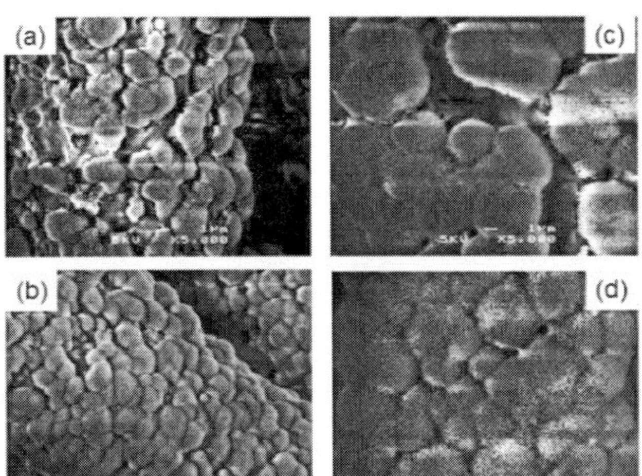

Figure 7. SEM images of the surfaces of (a,c) SiPVH30 and (b,d) SiPVH60 after (a,b) 1 days and (c,d) 3 days of soaking in 1.5SBF.

Bilayered Fibremats

Two Different Fibremats

PLLA fibremats were prepared by an electrospinning method and pressed at 15 MPa at room temperature to reduce the sizes of the spaces between the fibres for the purpose of preventing the intrusion of soft tissue. The pressed PLLA fibremat was incorporated with as-prepared SiPVH60 fibremat by hot-pressing. The pressed PLLA fibremat with ~100 μm in thickness was placed on top of the SiPVH60 one, and then pressed at ~ 0.2 MPa with a stainless steel mesh with an opening size of ~ 400 μm which was heated at 150 °C for 10 s, resulting in a bilayeredfibremat. The bilayeredfibremat was soaked in 1.5SBF.

Figure 8 shows the cross-section and the surface of the bilayeredfibremat. The fibres were fused at the hot-pressed regions. There were no significant changes in the shapes of fibres and the spaces between the fibres in the area that was not hot-pressed. No coating was observed on the PLLA fibres even after soaking in 1.5SBF, although SiPVH60 fibres were coated with cauliflower-like HCA. The resulting bilayeredfibremat was flexible and also strong enough to avoid detachment and damage when pulled by hand.

Cell Attachment

Mouse osteoblast-like cells (MC3T3-E1 cells) were seeded onto the bilayeredfibremat (SiPVH60 surface) placed in a 24-well plate at a density of 30,000 cells/well and incubated for 5 days. Alpha minimum essential medium (αMEM) containing 10 % fetal bovine serum (FBS) was used as a culture medium. After culturing, the cells were incubated for Alexa Fluor 488 Phalloidin and viewed with a fluorescence microscope.

Figure 9 shows the cell morphology on the bilayeredfibremat. An image of the bilayeredfibremat after culturing which was taken with a light microscope was superimposed on the cell image taken with a fluorescence microscope to find where in the fibremat the cells attached. The cells were observed around the bonded portion, which was formed by the hot-pressing. The fibre structure with small-sized spaces, that is, the bonded portion might not be favoured for the cell attachment in this sample.

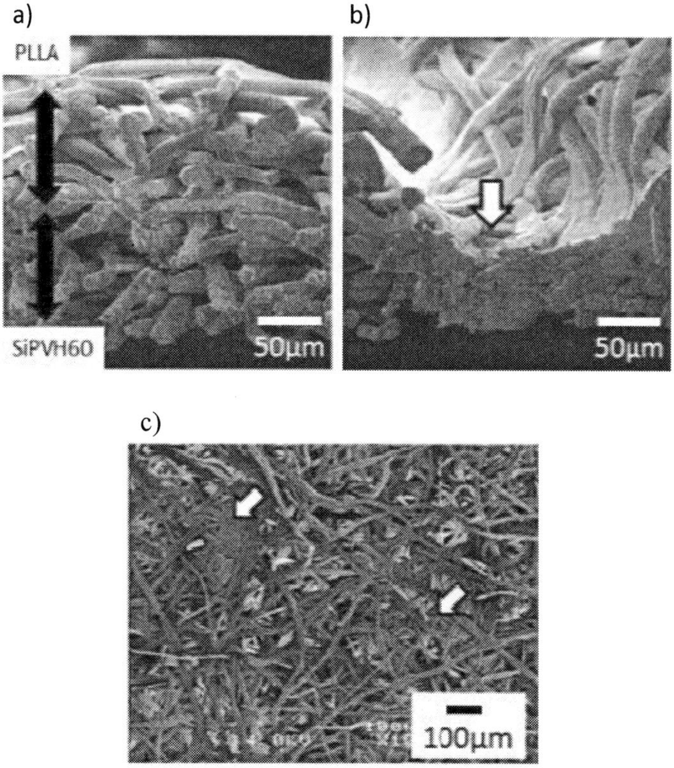

Figure 8. SEM images of the cross-sections (a) in an unbonded portion and (b) around the bonded portion and (c) the surfaces of SiPVH60 after hot-pressing. Black arrows indicate the area of PLLA or SiPVH60.White arrows indicate the bonded portion.

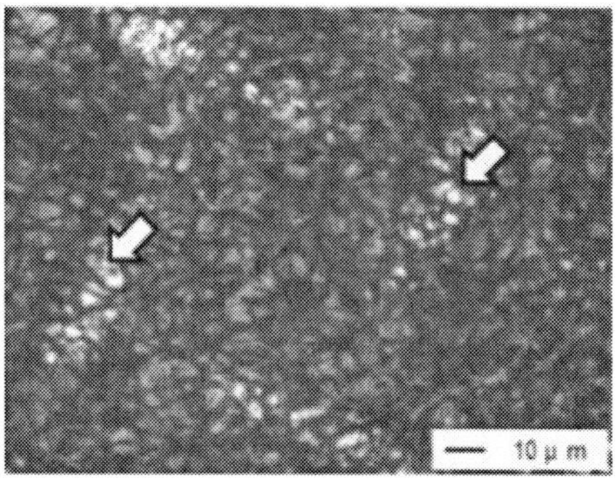

Figure 9. Morphology of osteoblast-like cells (MC3T3-E1 cells) on the bilayeredfibremat after 5 days-culturing. Green indicates the cells. Arrows indicate the bonded portion.

Tissue Compatibility

The tissue compatibility and bone-forming ability of the bilayeredfibremat was evaluated by *in vivo* tests using rabbits. An 8-mm diameter hole was drilled into the front midline of the animal's calvaria using a bone cutter and then covered with the bilayeredfibremat. An 8-mm diameter hole in bone is generally not regenerated if remained to be done. The bilayeredfibremat was fixed by putting between skin and bone. The bilayeredfibremat was implanted with SiPVH60 fibremat on the side in contact with the hole and dense PLLA one on the side in contact with the skin.

Figure 10 shows the histology of *in vivo* response to the bilayeredfibremat and the new bone formation at the center of the holes. Villanueva Goldner stain showed that mineralized tissue formation started in the bonded portion formed by hot-pressing. Mineralized tissue was formed over almost the entire area of implanted SiPVH60 fibremat after 12 weeks. There was no tissue inflammation observed by histology. The bone formation started from the inside of SiPVH60 fibremat and not from the edge of the hole. In our preliminary experiment, a poly(lactide-*co*-glycolide) fibremat was implanted in the same position for 12 weeks and new bone was not to be formed in/on the mesh at the center of an 8-mm hole. The bilayeredfibremat is expected to show not only the function of the barrier membrane to control the intrusion of the soft tissue but also importantly the function of a scaffold with three dimensional structure, cell activation and enhanced bone-forming ability.

a) b)

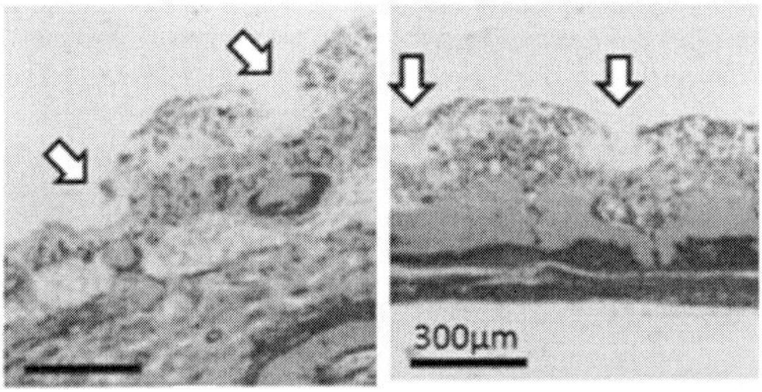

Figure 10. Histology of in vivo response to the bilayeredfibremats after (a) 4 weeks and (b) 12 weeks. Villanueva-Goldner staining shows mineralized tissue formed in the fibremat at the center of the holes. Green indicates mineralized tissues. Arrows indicate the bonded portion.

CONCLUSION

The PLLA-based composite fibremat releasing silicon and calcium species was prepared by an electrospinning method. The results showed here suggested the fibremat as an attractive candidate for GBR membrane with an ability enhancing bone formation in body. SiV particles, which are source of the species, were successfully incorporated with a PLLA matrix. The mechanical properties such as tensile strength and elongation of the fibremat strongly related to the content of SiV in the composite. It is necessary to find an appropriate weight ratio of polymer matrix: SiV particle to get a fibremat showing flexibility, strong, and releasing ability of silicon and calcium species enough for enhancement of bone regeneration. We suggest a new technique for fabricating such fibremat; incorporating two different fibremat, PLLA fibremat and SiPVH60 one, by hot-pressing with a stainless mesh. The bonded portion formed by hot-pressing played an important role both in the cell attachment *in vitro* and bone formation *in vivo*, particularly it was a very interesting to find the mineralized tissue formation started around the bonded portion.

ACKNOWLEDGMENTS

This work was supported in part by Grant-in-Aids for Scientific Research from Japan Society for Promotion of Science (JSPS) (20390497) and from The Ministry of Education, Culture, Sports, Science and Technology (MEXT) for Young Scientists (21700487) and by JSPS International Training Program (A program for developing young researchers in next-generation ceramic science and establishing an international network).The authors would also like to thank Dr. Yoshio Ota of Yabashi Industries, Co. Ltd., Dr. Takashi Wakita of Yamahachi Dental Manufacturing Co., Dr. Sen Lin, Miss Kie Fujikura and Mr. Shinya Yamada for supporting this work.

REFERENCES

[1] Hench, L.L. *Journal of the American Ceramic Society*,81 1705-1727 (1998).

[2] Kim, S.-S., Sun Park, M., Jeon, O., Yong Choi, C., and Kim, B.-S. *Biomaterials*,27 1399-1409 (2006).

[3] Zhang, K., Wang, Y., Hillmyer, M.A., and Francis, L.F. *Biomaterials*,25 2489-2500 (2004).

[4] Kasuga, T., Maeda, H., Kato, K., Nogami, M., Hata, K.-I., and Ueda, M. *Biomaterials*,24 3247-3253 (2003).

[5] Bramwell, V.W., Eyles, J.E. and Alpar, H.O. *Advanced Drug Delivery Reviews,*57 1247-1265 (2005).

[6] Santerre, J.P., Woodhouse, K., Laroche, G., and Labow, R.S. *Biomaterials*,26 7457-7470 (2005).

[7] Jones, J.R., Tsigkou, O., Coates, E.E., Stevens, M.M., Polak, J.M., and Hench, L.L. *Biomaterials*,28 1653-1663 (2007).

[8] Yuan, H., Kurashina, K., De Bruijn, J.D., Li, Y., De Groot, K., and Zhang, X. *Biomaterials*,20 1799-1806 (1999).

[9] Sill, T.J. and Von Recum, H.A. *Biomaterials*,29 1989-2006 (2008).

[10] Li, W.J., Laurencin, C.T., Caterson, E.J., Tuan, R.S., and Ko, F.K. *Journal of Biomedical Materials Research*,60 613-621 (2002).

[11] Xin, X., Hussain, M. and Mao, J.J. *Biomaterials*,28 316-325 (2007).

[12] Badami, A.S., Kreke, M.R., Thompson, M.S., Riffle, J.S., and Goldstein, A.S. *Biomaterials*,27 596-606 (2006).

[13] Yang, F., Murugan, R., Wang, S., and Ramakrishna, S. *Biomaterials*,26 2603-2610 (2005).

[14] Zeng, J., Xu, X., Chen, X., Liang, Q., Bian, X., Yang, L., and Jing, X. *Journal of Controlled Release*,92 227-231 (2003).

[15] Hoppe, A., Güldal, N.S. and Boccaccini, A.R. *Biomaterials*,32 2757-2774 (2011).

[16] Carlisle, E.M. *Ciba Foundation symposium*,121 123-139 (1986).

[17] Xynos, I.D., Edgar, A.J., Buttery, L.D.K., Hench, L.L., and Polak, J.M. *Biochemical and Biophysical Research Communications*,276 461-465 (2000).

[18] Xynos, I.D., Edgar, A.J., Buttery, L.D.K., Hench, L.L., and Polak, J.M. *Journal of Biomedical Materials Research*,55 151-157 (2001).

[19] Gough, J.E., Jones, J.R. and Hench, L.L. *Biomaterials*,25 2039-2046 (2004).

[20] Reffitt, D.M., Ogston, N., Jugdaohsingh, R., Cheung, H.F.J., Evans, B.A.J., Thompson, R.P.H., Powell, J.J., and Hampson, G.N. *Bone*,32 127-135 (2003).
[21] Obata, A., Tokuda, S. and Kasuga, T. *Acta Biomaterialia*,5 57-62 (2009).
[22] Obata, A., Hotta, T., Wakita, T., Ota, Y., and Kasuga, T. *Acta Biomaterialia*,6 1248-1257 (2010).
[23] Wakita, T., Obata, A. and Kasuga, T. *Materials Transactions*,50 1737-1741 (2009).
[24] Wakita, T., Obata, A., Poologasundarampillai, G., Jones, J.R., and Kasuga, T. *Composites Science and Technology*,70 1889-1893 (2010).
[25] Wakita, T., Nakamura, J., Ota, Y., Obata, A., Kasuga, T., and Ban, S. *Dental Materials Journal*,30 232-238 (2011).
[26] Kokubo, T., Kushitani, H., Sakka, S., Kitsugi, T., and Yamamuro, T. *Journal of Biomedical Materials Research*,24 721-734 (1990).

In: Lactic Acid
Editors: A. Migallón and R. Ruseckaite

ISBN: 978-1-62081-643-1

Chapter 4

PROPERTIES OF LACTIC ACID BASED POLYMER AND ITS NANOCOMPOSITES

Kikku Fukushima*
Polytechnic of Turin, branch of Alessandria,
Alessandria, Italy

ABSTRACT

Lactic acid (2-hydroxypropanoic acid) is one of the smallest optically active molecules which can be either of L(+) or D(-) stereoisomer. This is produced by animals, plants, microorganisms in the nature and can also be derived from intermediates with an origin in renewable materials or from chemicals derived from coal or oil. The research on lactic acid based polymers [poly(lactic acid)-PLA-] intended for medical applications has been accelerated since 1960 and during the two last decades an increasing activity on large scale industrial lactic acid based polymers for environmental uses has been also noticed, probably due to the recent need for the urgent development of biodegradable materials that would not involve the use of toxic components in their manufacture and could be degraded in relatively short times to environmentally friendly products.

The main limitations of this polymer towards its wider industrial application are its poor thermal and mechanical resistance and limited fire retardance properties, restricting its access to industrial sectors, such as

* E-mail address: kikku81@gmail.com.

transport applications, in which its use would be justified when biodegradability is required. Nevertheless, it has been found that the above drawbacks can be overcome by enhancing their thermal and mechanical properties through copolymerization, blending and filling techniques. Indeed, the addition of nano-sized fillers could potentially confer multifunctional enabling properties to this polymer. In this way, nanoparticles in biodegradable polymers have a strong promise in designing eco-friendly nanocomposites for several applications. A fairly new area of composites has emerged in which the reinforcing material has the dimensions in nanometric scale, such composites are called nanocomposites. It has been proved that these nanocomposites can significantly improve thermal and mechanical properties of several polymers even with a very low level of nanoparticle addition due to their nanoscale dispersion, so that the reinforcement efficiency of nanofillers corresponds to that of 40-50% loading of classical micro-fillers.

This chapter was addressed to review the effect of addition and co-addition of different nanoparticles on the thermal, mechanical and fire retardant properties of PLA. As general result, potential bio-based compositions for industrial applications were found with the mixing of PLA and these used nanoparticles, achieving important enhancements in the rigidity and in fire retardance of the PLA matrix. Even more, it was found that the co-addition of thenanoparticles in the polymer matrix resulted in the occurrence of synergistic interactions between nanoparticles on the above properties, therefore regarding these compositions as potentially interesting materials for the enhancement of PLA properties for transport applications, electric and electronic engineering materials.

Keywords: PLA, nanocomposites, clays, graphite, industrial applications

INTRODUCTION

Lactic acid is an organic acid found in many products of natural origin. The first report on isolation of lactic acid from milk can be found as early as 1780 and the solidification by self-esterification some years later. Then, the dimerization of polycondensated lactic acid into lactide and the ring-opening polymerization thereof was reported in 1932. However, at that time, the polymers based on lactyl units were found to be unstable at humid conditions and the use of these was not considered as relevant before in 1960s, when the advantages from their use in medical applications became apparent.

Lactic acid (2-hydroxypropanoic acid) is one of the smallest optically active molecules, which can be either of L(+) or D(-) stereoisomer [1]. This is produced by animals, plants, microorganisms in the nature and can be also derived from intermediates with an origin in renewable materials or from chemicals derived from coal or oil. The synthesis of lactic acid comes from pyruvate under oxygen limiting conditions in a lactic acid dehydrogenase catalyzed reaction, in which the stereoisomeric ratio of D- and L-lactic acid is determined by specificity of the dehydrogenase [1]. Fermentative production of lactic acid, which can be performed in batch or in continuous processes, is preferably made by the group of lactic acid bacteria capable of converting hexoses into lactic acid.

The research on lactic acid based polymers (PLA) intended for medical and large industrial applications has been accelerated during the two last decades, especially for industrial applications considering that there is an urgent need for the development of industrial biodegradable materials that would not involve the use of toxic components in their manufacture and could be degraded in relatively short times to environmentally friendly products; as well as an alternative to traditional commodity plastics due to an increased environmental concern related to the use of non-biodegradable polymers in everyday life [2,3].

In general, the melt processing and physical properties of PLA are similar to those of conventional polymers; indeed in many aspects the properties of PLA lie between those of crystal polystyrene (PS) and polyethylene terephthalate (PET), making possible its use as a commodity polymer for general applications. Even more, nowadays, this polymer has acquired one of the most important positions on the market of biodegradable polymers due to its possible fields of application including a wide variety of biomedical products (fracture fixation devices, bioresorbable surgical sutures, dental implants, bone screws, plates and controlled drug delivery), food packaging, films for agro-industry, fibres production and, as one of the last tendencies, as composites for technical applications, such as electronic and electrical devices, mechanical or automotive parts [4-6]. However, it has been reported that some of PLA properties are often not good enough for several specific industrial applications [5,7], such as transport which requires a high level of mechanical and flame retardant properties.

The above drawbacks could be overcome by addition of nano-sized particles [7]. Various nano-reinforcements in polymer matrices are being developed but the most extensively studied polymer nanocomposites so far, are those based on layered silicate clays, due to their availability, low cost and enhancements that they can induce in mechanical, thermal, barrier and flame retardant properties of the polymer matrix, in comparison to more conventional microcomposites[7-10].

In this way, Ogata *et al.* [11] reported the preparation of PLA/organically modified layered silicate nanocomposites by dissolving the polymer in hot chloroform in the presence of dimethyl distearyl ammonium modified montmorillonite. The morphological results showed that the silicate layers could not be intercalated in the PLA matrix, as the clay existed in the form of tactoids. On the other hand,Bandyopadhyay*et al.* [12] reported the preparation of intercalated PLA/organically modified layered silicate nanocomposites by using the solvent casting method, with much improved mechanical and thermal properties.

Sinha Ray *et al.* [13-15] used melt technique for the preparation of intercalated PLA/layered silicate nanocomposites, obtaining that the silicate layers of the clay were intercalated and randomly distributed in the PLA matrix. Parallel, Fukushima *et al.* [16,17] could also obtain PLA based nanocomposites based on montmorillonite with a very high exfoliation and intercalation level of this clay with the polymer matrix, resulting in high thermal and mechanical enhancements in the PLA matrix.

Concerning fire retardancy properties of clay-based polymer nanocomposites, it has been reported that a relatively low quantity of nanoclays in the polymer matrix can create a protective layer during combustion [18]. Moreover, heat transfer promotes thermal decomposition of the organomodifier and the creation of strongly protonic catalytic sites onto the clay surface, which can catalyze the formation of a stable char residue [18]. Therefore, accumulation of the clay on the surface of the material can act as a protective barrier that limits heat transfer to the material and hence reducing the volatilization of combustible degradation products, which feed the flame and the diffusion of oxygen into the material.

As alternative to clays, graphite is also among the leading nanoscale fillers in research and development of commercial products [19,20]. Graphite is a layered carbon-based material comparable to clays, made however of neutral graphene layers whereas clays used for nanocomposites preparation are negatively charged because of isomorphous substitution within the layers. In clays, the charge on the layers favors intercalation of organic cations, which increase the interlayer space of the clay and its affinity to polymer chains. In the case of graphite, in which the carbon atoms are bonded covalently in a hexagonal arrangement in the same plane and only much weaker van der Waals forces are acting between the successive adjacent planes, since the forces between graphite nanosheets are relatively weak, it is possible for a wide range of atoms, molecules and ions to intercalate between layers. Intercalation compounds such as the expandable graphite, which provide a source for polymers nanocomposite formation [21], are in general prepared by graphite oxidation treatments, commonly in the presence of concentrated H_2SO_4 and HNO_3 acids. The synthesis of expanded graphite is then obtained by rapid expansion and exfoliation of expandable graphite in a furnace at high temperature [21,22].

The exfoliated graphite nanoplatelets (graphene), consisting of a large number of delaminated graphite sheets, combine the low price and layered structure of clays with the superior thermal and electrical properties of carbon nanotubes (CNTs), thus can be an effective alternative to both clays and nanotubes. Thisnanofiller provides excellent competitive functional properties to polymers for different applications [22-24], especially in electromagnetic interference shields, thermal conductors and in the automotive industry (fuel injection and anticorrosion systems, fuel tank inlet, electrostatically paintable parts, etc.).

In recent years, polymer based nanocomposites reinforced with expanded graphite have shown substantial improvements in mechanical, electrical conductivity and barrier properties in several polymers, such as poly(methyl methacrylate) (PMMA), polyamide-6 (PA-6), polyethylene (PE), polystyrene (PS) [21,24-27]. The reason for this is that graphene sheets/layers characterized by high modulus, could be separated down to a nanometer thickness, with high aspect ratio. Therefore, the dispersion of such nanosheets in a matrix can play a key role in the improvement of both physical and mechanical properties of the resultant nanocomposites [21,28,29].

In the literature there are only few reports regarding the effect of expanded graphite on the thermal stability and flame retardance of polymers. Fawn *et al.* [30] studied the mechanical and fire retardant properties of PA-6/graphite nanocomposites by melt-blending, using virgin graphite, expandable graphite and expanded graphite, obtaining that when they compared to PA-6/montmorillonite clay nanocomposites of similar filler dispersion, the PA-6/graphite nanocomposites showed higher improvements in flame retardant properties, especially by using the expandable graphite, but not as good thermo-mechanical properties as the PA-6/clay nanocomposite equivalents.

On the other hand, graphite is also recognized for its self-lubricating and dry lubricating behavior, and it is expected that for engineering utilizations its addition into a biopolymer of interest for further developments, such as PLA, could enhance simultaneously its wear resistance and its electrical properties.

In this regard, in response to the demand for extending PLA applications range it has been recently demonstrated [17,31] that PLA can be melt-blended with expanded graphite using different techniques of addition, and thus obtaining materials characterized by improved properties (rigidity, crystallization, moderate increase of thermal stability and flame retardant properties). To open the way to products with multifunctional properties, it is believed that the performances of PLA-graphite nanocomposites can be further improved by co-addition of specific nanofillers or additives. Thus, considering this background, the aim of the present chapter was to review the effect of addition/co-addition of expanded graphite and of a selected organically modified clay on the mechanical and fire retardant properties of PLA, in order to evaluate the possible use of new bio-based lactic acid based products as electric and electronic engineering materials, as well as in automotive parts.

EXPERIMENTAL

Materials and Methods

Poly(lactic acid) (PLA, 4032D) was a commercial grade supplied by Nature Works (US). Two nanoparticles were used: an organically modified layered silicate supplied by Southern Clay (US) -CLOISITE 30B- and an expanded graphite supplied by Timcal Belgium S.A. -TIMREX BNB90-. The characteristics of the nanoparticles are listed in Tables 1 and 2.

Table 1. Characteristics of montmorillonite

Commercial name	Modifier chemical structure	Particlesize (μm)	Notation in text
CLOISITE 30B	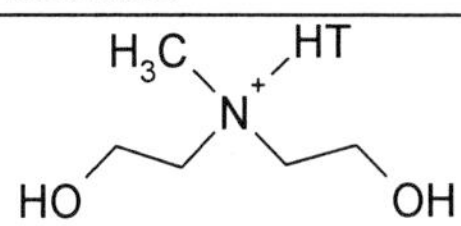 *bis*-(2-hydroxyethyl) methyl hydrogenated tallow ammonium)*	10% < 2 50% < 6 90% < 13	CLO

*Organic fraction about 30 wt-% determined by Thermogravimetric Analysis (TGA).

Table 2. Characteristics of expanded graphite

Commercial name	Particle Size (μm)	Notation in text
TIMREX BNB90	10% < 10 50% < 36 90% < 85	EG

It is important to highlight that, as reported elsewhere [32], the production of the expanded graphite used in this chapter consisted in the following procedure: the natural graphite flakes were treated at elevated temperatures with mixtures of sulphuric acid (H_2SO_4) and hydrogen peroxide (H_2O_2) or ammonium persulfate ($NH_4S_2O_8$). Under these conditions graphite was oxidized and at the same time sulphate anions were inserted between the graphite sheets. The resulting graphite salt was isolated by filtration, washed up to some extent and dried. The expansion of the graphite salt was then performed by brief exposure to temperatures above 700°C, which caused the decomposition of the acid residues between the graphite layers to gaseous products.

Prior to the mixing step, PLA, CLO and EG were dried at 100ºC under vacuum for 4 hours. Several PLA based nanocomposites were prepared according to the compositions showed in Table 3.

The nanocomposites were obtained by mixing PLA pellets with CLO and/or EG in an internal mixer at 190°C. The mixing was performed at two different rotor speeds: 30 rpm in the loading step (3 min) and 70 rpm during mixing (7 min). The batch was extracted from the mixing chamber manually, followed by cooling under air to room temperature. Neat PLA, binary and ternary compositions (see Table 3) were produced using similar conditions of preparation. Plates were obtained by compression molding at 190ºC using a hydraulic press. The materials were first pressed at low pressure for 200 s (2 degassing cycles), followed by a high pressure cycle at 150 bars for 150 s step followed by cooling at 50 bars for 300 s. Specimens for burning tests were also obtained from compression moulded plates, by using a sample mould in accordance with the ASTM D3801/UL94V standard (125 mm x 13 mm x ~3.3 mm). For tensile tests the samples (specimens type ISO½ with thickness of 1.5 mm) were obtained using a DSM Micro-Injection Moulding Machine (3.5 cm^3) at processing temperature of 200°C and mould temperature of 80°C.

Table 3. Content of PLA, CLO and EG used for the preparation of nanocomposites

Composition	PLA (wt-%)	CLO*(wt-%)	EG (wt-%)	Notation in text
1	100	-	-	PLA
2	97	-	3	PLA-3EG
3	94	-	6	PLA-6EG
4	91	-	9	PLA-9EG
5	97	3	-	PLA-3CLO
6	94	3	3	PLA-3CLO-3EG
7	91	3	6	PLA-3CLO-6EG
8	88	3	9	PLA-3CLO-9EG

*The percentages given as wt-% are related to the amount of inorganic phase in the resulting nanocomposites.

Morphological characterizations were performed on specimens (25 mm diameter x 1.5 mm thickness) prepared in similar conditions by injection moulding.

Characterization Techniques

Wide Angle X-Ray scattering analysis was performed on a Siemens D5000 diffractometer using Cu Ka radiation (wavelength, 1.540 Å) at room temperature with scanning rate 2°min^{-1}. Transmission electron micrographs were obtained with a Philips CM100 apparatus using an accelerator voltage of 100 kV. The samples (70–80 nm thick) were prepared with a Reichert Jung Ultracut 3E FC4E ultracryomicrotome by cutting at -130°C.Tensile testing measurements were performed by using a Lloyd LR 10K tensile bench in accordance to the ASTM D638-02a at a speed rate of 1 mm min^{-1}, preload of 10 N, using a distance of 58 mm between grips. All tests were carried out by using at least 7 samples previously conditioned for 48 hours at 20 ±2°C under a relative humidity of 50 ±3%.

The vertical burning test UL94 V (Underwriters Laboratory, applicable as ASTM D3801 and ISO 1210) covers a scale laboratory procedure for determining vertical burning characteristics of solid plastic materials for a particular application, using a flame applied to the base of specimens held in a vertical position. The procedure consists of subjecting a set of 5 specimens (125 mm x 13 mm x 3-13 mm) to a standard test flame [17]. The afterflame times t_1 and t_2 are noted after a first and second flame application respectively [17], as well as the afterglow time t_3 (time for which the material continues to glow after the ignition source has been removed and/or cessation of flaming) and information on whether any particles fall from the specimen and, if so, whether they ignite the cotton. This standard also provides a classification system to characterize the vertical burning behavior of the test samples as described in Table 4.

Additionally, in order to obtain additional information on the fire burning behaviour of samples, each specimen was marked at 25 mm and 100 mm from one end of the test specimen. When the flame passes the 25 mm and 100 mm mark, the burning time was recorded and the vertical BR (mm/min) was calculated. At the same time, Scanning Electron Microscopy (SEM) was carried out on the fracture surface of residual materials after UL94 V test, by using a SEM apparatus LEO 1450 VP equipped with energy dispersive spectroscopy (EDS).

Table 4. Specifications of materials classification according to the Standard ASTM D 3801

Criteria conditions	V-0	V-1	V-2
Afterflame time for each individual specimen (t_1 or t_2)	≤ 10 s	≤ 30 s	≤ 30 s
Total afterflame time for each specimen set ($t_1 + t_2$ for the five specimens)	≤ 50 s	≤ 250 s	≤ 250 s
Afterflame plus afterglow time for each individual specimen after the second flame application ($t_2 + t_3$)	≤ 30 s	≤ 60 s	≤ 60 s
Afterflame of any specimen up to the holding clamp	NO	NO	NO
Cotton indicator ignited by flaming particles or drops	NO	NO	YES

RESULTSAND DISCUSSIONS

Morphological Analysis

Figure 1a represents the Wide angle X-ray scattering (WAXS) patterns of PLA-CLO nanocomposites as well as of pristine clay. CLO was characterized by a single diffraction peak at $2\theta \approx 4.7°$ corresponding to the basal reflection (001) and accounting 1.9 nm of interlayer distance [17].

As expected, according to the results of previous studies reported elsewhere [14,17,33,34], the addition of CLO revealed a good interaction of this clay with the PLA matrix. Indeed, a shift of the clay diffraction peak to lower angle was observed for the binary and ternary systems (Figure 1a), corresponding to an increase of the interlayer distance of about 1.9 nm, and mainly attributed to the formation of intercalated nanocomposites. On the other hand, it was possible to verify that in ternary compositions graphite sheets favoured clay exfoliation since the height of the diffraction peak associated to the intercalation of CLO decreased with increasing graphite content, especially upon addition of 9% EG.

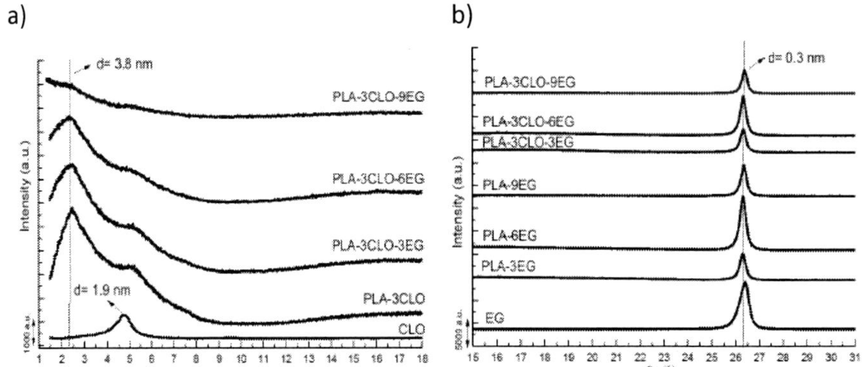

Figure 1. WAXS patterns of PLA nanocomposites compared to pristine nanofillers: (a) CLO and (b) EG.

Concerning to the EG composites, pristine expanded graphite was characterized by a single diffraction peak at $2\theta \approx 26.4°$ corresponding to the characteristic peak of pure graphite with a basal reflection (002) [17,21,19,30], and accounting 0.3 nm interlayer distance (Figure 1b) between graphene sheets in the layers stacks of expanded graphite.

The WAXS patterns of EG based materials nanocomposites did not show signs of intercalation of PLA between graphene sheets, likely due to the insufficient interlayer distance which is considerably lower as compared to CLO [17]. In fact, the presence of the sharp peak at 26.4° (see Figure 1b) confirmed not only the presence of pure graphite based on stacks of parallel grapheme sheets, but also the fact that the melt-compounding applied here is not able to exfoliate the graphene layers and some sheets still existed in the aggregate form [17,31].

Several PLA nanocomposites obtained by addition and co-addition of nanofillers were analyzed through TEM analysis. TEM patterns of PLA-3CLO showed a high distribution and dispersion level of the organoclay in the polymer matrix (Figure 2a). Even more, single dispersed exfoliated layers and thin intercalated stacks of layers were confirmed in a previous work [17] and visible into the PLA matrix in Figure 2a, thus the structure of this nanocomposite was characterized to be exfoliated-intercalated.

The addition of 6% EG into the PLA matrix resulted innanocomposites in which the dispersion of the nanoparticles was considered good enough knowing the initial particle/aggregate size (d_{50} = 36 microns [17,31]) and also the fragility of the EG layers that are broken down during melt-blending [31].Indeed, bundles of hundreds of nanometres or more, consisting in graphite platelets (Figure 2b) as well as individual graphite nanosheets (shown in Figure 2b) were observed throughout the polymer matrix. Moreover, the coexistence of single dispersed graphite layers and of 'tactoids' consisting of various layers (in agreement with WAXS analysis) obtained under moderate shear during melt blending revealed good interactions of these graphite particles with the PLA.A similar graphite dispersion was obtained upon addition of 3% and 9 % of EG.

Concerning the ternary compositions, TEM micrographs of PLA-3CLO-6EG showed a good dispersion obtained by co-addition of these two nanoparticles (Figure3a). Single clay layers, bundles of graphite platelets and well-dispersed graphite layers into the PLA matrix were observed in different magnification images (Figure 3a, 3b).

a) b)

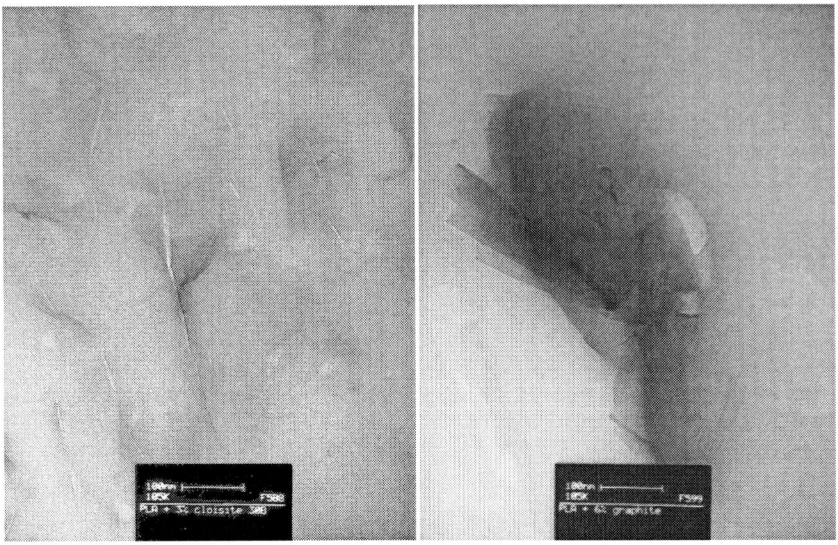

Figure 2. TEM patterns of (a) PLA-3CLO and (b) PLA - 6EG at 105 kX of magnification.

In parallel, the higher CLO and EG dispersion reported in these ternary compositions as compared to that obtain in the corresponding binary ones [17] confirmed that the presence of graphite further improved dispersion of the clay (consistent with the WAXS analysis, Figure 1a) and also suggests that the clay played a similar role towards graphite. Similar dispersion levels were achieved for PLA-3CLO-3EG and PLA-3CLO-9EG.

From WAXS and TEM results it was possible to assume that the dispersion and interaction of the CLO in PLA were better than those obtained with EG in binary and ternary compositions, and attributed to the chemical affinity between PLA and CLO originated from hydrogen bonding between the carbonyl groups of PLA and the hydroxyl groups of the clay surface and organic modifier of CLO [10,17,31,35]. However, it is also important to underline that a good dispersion of expanded graphite platelets was also achieved for all materials containing EG, and associated to the presence of some functional groups such as hydroxyl, carboxyl, carbonyl, etc., in the graphite platelets [17,21,36]. These functional groups could highly promote hydrogen bonding between the polyester matrix and graphite particles.

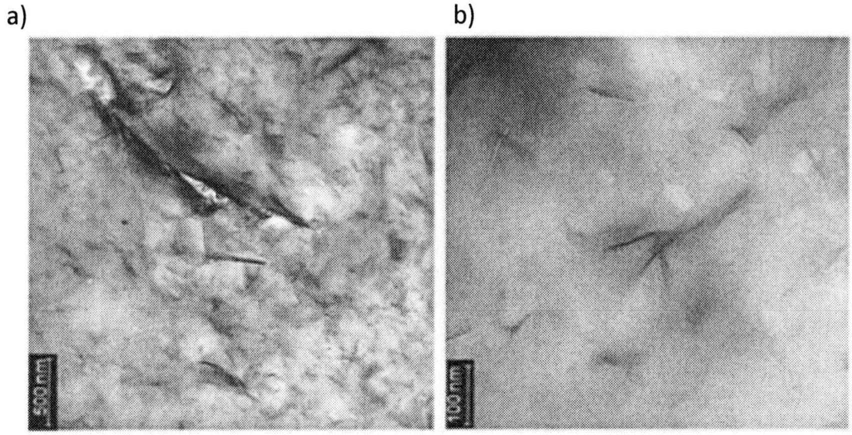

"Reprinted from Effect of expanded graphite/layered-silicate clay on thermal, mechanical and fire retardant properties of poly(lactic acid), 95/10,Kikku Fukushima, Marius Murariu, Giovanni Camino, Philippe Dubois, pages 1063-1074, Copyright (2010), with permission from Elsevier."

Figure 3. TEM patterns of PLA-3CLO-6EG at (a) 24 kX and (b) 100 kX of magnification.

Mechanical Properties

Table 5 presents the reported tensile properties of neat PLA and its nanocomposites tested at room temperature [18]. For most of the composites, the maximum tensile strength remained practically constant as compared to neat PLA. For all specimens containing EG, the nominal strain at break of PLA tended to decrease gradually by increasing the percentage of expanded graphite, whereas the PLA-3CLO nanocomposite presented an increase of this parameter with respect to the neat polymer matrix, and assumed to be due to a slight plasticizer effect conferred by the presence of organo-modifier in the clay [17].

On the other hand, a high rigidity is another key-property required for mechanical or engineering applications, and the addition of nanofillers with high modulus should lead to a significant reinforcing effect and the increase of rigidity of PLA [31]. Indeed, significant improvements were reported [17] in the rigidity of PLA upon addition of CLO and EG: at low loading, *i.e.*, 3 wt-%, the addition of CLO and EG into PLA led to a similar increase of the Young's modulus by *ca.* 20%, and these increases were associated to the good dispersion and reinforcement effect of both nanoparticles in the polymer matrix. Moreover, binary systems based on expanded graphite showed increases of the Young's modulus directly correlated to the graphite percentage, for example, the addition of 6% and 9% EG into PLA led to increases of Young's modulus by *ca.* 30% and 50%, respectively, with respect to neat PLA.

Furthermore, it is important to underline that the highest improvements in the rigidity were reported for the ternary systems (see Table 5) [17], achieving notable increases of the Young's modulus of 50% for PLA-3CLO-3EG, 54% for PLA-3CLO-6EG and 82% for PLA-3CLO-9EG. These important mechanical improvements in rigidity were most likely due to co-reinforcement effect of both nanofillers and to their additive/favourable interactions with PLA.

Table 5. Mechanical properties of neat PLA andPLA nanocomposites (standard deviation in parenthesis)

Sample	Maximum tensile strength(MPa)	Nominal strain at break(%)	Young's modulus (MPa)
PLA	63 (±2)	3.5 (±0.1)	2800 (±120)
PLA-3CLO	64 (±1)	4.7 (±0.6)	3400 (±100)
PLA-3EG	58 (±1)	3.4 (±0.5)	3300 (±180)
PLA-6EG	57 (±1)	3.0 (±0.4)	3700 (±180)
PLA-9EG	57 (±1)	2.5 (±0.1)	4200 (±130)
PLA-3CLO-3EG	64 (±1)	2.6 (±0.1)	4200 (±280)
PLA-3CLO-6EG	60 (±1)	2.4 (±0.1)	4300 (±300)
PLA-3CLO-9EG	63 (±1)	2.1 (±0.1)	5100 (±400)

"Data extracted from Effect of expanded graphite/layered-silicate clay on thermal, mechanical and fire retardant properties of poly(lactic acid), 95/10,Kikku Fukushima, Marius Murariu, Giovanni Camino, Philippe Dubois, pages 1063-1074, Copyright (2010), with permission from Elsevier."

Fire Retardant Properties

Fire hazards result from the combination of different factors (ignition, amount of heat released on burning, rate of heat release, flame spreading, etc.) and different procedures can be used for evaluation of the flame retardant properties of polymer materials [31]. Some nanofillers (carbon nanotubes, graphite, clays) have recently received much attention and are considered as potential additives for their positive impact on principal thermo-mechanical properties and also in terms of flame retardancy. Generally these nanofillers lead to the formation of a protective thermally stable surface layer during combustion, which limits the heat transfer from the flame to the substrate and mass transfer from the substrate to the flame. Thus, the overall rate of flame feeding by combustible products from polymer pyrolysis and thermo-oxidation is decreased [31].

The vertical burning test UL94 V is one of the most commonly used tests for measuring the ignitability and flame spread of vertical bulk materials exposed to a small flame. Generally, according to the results shown in literature, the polymer nanocomposites can be classified according to this standard especially when the nanofillers are associated to traditional flame retardant additives. For producing new polymer compositions characterized by improved flame retardant properties, some research works have been dedicated to the study of the combinations between polymer layered-silicate nanocomposites and conventional flame retardants [17,37].The general approach in these studies starts with a conventional flame-retarded polymer formulation, which already shows good flammability properties. (*i.e.*, a V0 rating in the UL94 flammability test). Addition of nano-dispersed layered-silicate allows a significant portion of the conventional flame retardants to be removed from the formulation.

Table 6 shows the main results reported for the studied PLA and nanocomposites after vertical burning test [17]. Neat PLA presented some fire retardance, with the occurrence of flame dripping able to remove heat from the surface, and explained by the removal of the inflamed polymer material from pyrolysis zone by dripping with effect in decreasing the fuel support rate [17]. This dripping allowed PLA to achieve a V-2 classification and to leave a residual polymer mass of *ca.* 64% in the holding clamp, however, this dripping was also able to ignite the cotton limiting the use of PLA in electrical applications.

Because the PLA nanocomposites did not accomplish different criteria (the after flame time for individual specimen, total time of burning, etc.), they could not be classified according to the standard of UL94 V test [17].However, in spite of these results, some other selected aspects concerning their behavior during burning were interesting and reported in this chapter. The addition of CLO and EG in PLA (binary compositions) showed benefits during combustion, such as considerable decreases of the sample BR as compared to neat PLA (see Table 6). This phenomenon was related to the formation of a carbonized surface layer rich in silicate or graphite, able to protect the bulk

PLA from the heat source. While PLA-CLO nanocomposites preserved the original shape of specimens with partial fragmentation, PLA-EG nanocomposites were completely fragmented by burning and dripping [17]. These last materials wereburning with formation of an intumescent char with initial viscous consistency due to melting of PLA (see Table 6), and the volume of resulting carbonaceous char was in direct correlation with the percentage of EG nanofillers, similar observations were previously reported for the same samples during their horizontal burning tests [17].

Table 6. Results for PLA and nanocomposites after vertical burning test UL94 V (standard deviation reported in parenthesis)

Sample	Total burning time(s)	Mean BR (mm/min)	Flame dripping	Ignited cotton	Residual mass (%)	Observations
PLA[a]	94 (±5)	189 (±38)	Yes	Yes	64 (±4)	Intense dripping
PLA-3CLO*	291 (±7)	124 (±14)	Yes	Yes	26 (±1)	Burning with drips, charring
PLA-3EG*	298 (±9)	127 (±16)	Yes	Yes	31 (±5)	Burning with drips, charring
PLA-6EG*	336 (±7)	113 (±11)	Yes	Yes	31 (±5)	
PLA-9EG*	382 (±9)	110 (±8)	Yes	Yes	35 (±5)	
PLA-3CLO-3EG*	432 (±5)	95 (±5)	No	Yes	40 (±2)	Compact char**
PLA-3CLO-6EG*	475 (±7)	81 (±5)	No	No	55 (±1)	Compact char
PLA-3CLO-9EG*	862 (±12)	78 (±3)	No	No	58 (±5)	Compact char

[a] The BR of neat PLA was calculated as the required time when the flame passes a 25 mm and 55 mm mark for each individual specimen.

* These specimens showed only one afterflame time (t_1) recorded as flame impingement time after completed burning of the sample up to the holding clamp. The mean BR was calculated as the required time when the flame passes the 25 mm and 100 mm mark for each individual specimen.

Other remarks:

** When flame up to holding clamp the sample breaks in two pieces. Fallen piece ignites cotton.

"Data extracted from Effect of expanded graphite/layered-silicate clay on thermal, mechanical and fire retardant properties of poly(lactic acid), 95/10,Kikku Fukushima, Marius Murariu, Giovanni Camino, Philippe Dubois, pages 1063-1074, Copyright (2010), with permission from Elsevier."

a) b)

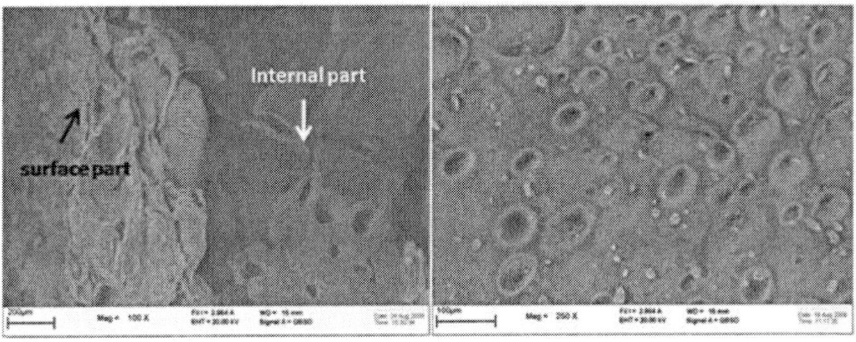

"Reprinted from Effect of expanded graphite/layered-silicate clay on thermal, mechanical and fire retardant properties of poly(lactic acid), 95/10,Kikku Fukushima, Marius Murariu, Giovanni Camino, Philippe Dubois, pages 1063-1074, Copyright (2010), with permission from Elsevier."

Figure 4. SEM micrographs of residual material from PLA-3CLO after UL94 V test: (a) in the sample fracture zone and (b) on the surface part from (a).

In these cases, a purely physical effect of the nanofiller was proposed and related to the formation of a ceramic surface layer by the accumulation of inorganic particles on the surface, acting as a physical barrier, shielding the underlying polymer from the radiated heat and limiting polymer ablation, thus reducing the amount of combustible volatile products available for burning in the gas phase and resulting in lower burning rate (BR). An additional catalytic chemical effect of the nanofillers has been suggested to explain formation of charring during combustion [38]. Indeed, the hypothesis concerning the formation of the carbonized ceramic protective surface layer was confirmed through SEM analysis on the fracture surface for these materials [17]. SEM micrographs obtained for PLA-3CLO residue are shown in Figure 4. The surface of PLA-3CLO residue (surface part in Figure 4a, Figure 4b) appeared as a continuous carbon layer rich in silicon and oxygen according to EDS analysis (Si: 19 wt-% as compared to 1.3 wt-% in original sample), explained by the accumulation of silicate layers at the sample surface. The higher Si content at the surface of the burnt specimen as compared to the bulk is also qualitatively appreciable by the lighter shade of gray for the surface when using a backscattered electron detector and due to the higher atomic weight of silicon Si compared to C (carbon) [17].

Regarding expanded graphite composites, SEM micrographs obtained on the fracture surface for PLA + 3EG residue fragments are shown in Figure 5. On the top layer (surface part in Figure 5a, Figure 5b) the accumulation of expanded graphite layers were observed whilst the bulk zone of the material (indicated as "internal part" in Figure 5a) presented a smooth continuous layer rich in carbon related to the underlying polymer protected from combustion. A similar trend was observed for binary compositions obtained by addition of 6% and 9% EG [17].

The formation of the carbonaceous surface layer rich in silicate or graphite was reported to be the main reason of the lower BR of these materials as compared to neat PLA. Indeed, Murariu*et al.*[31] found promising results on the flame retardant properties of PLA composites based on expanded graphite during their burning test by using a cone calorimeter with a defined heat flux of 35 kW m^{-2}, heat flux which corresponds to the common heat flux in mild fire scenario. They obtained that PLA-EG samples showed the beneficial role of graphite in the decrease of the maximum value of the rate of heat release (RHR), thus reducing the possibility of fast ignition and flame spreading. Even more, they also reported that for all formulations containing EG, by comparing to neat PLA, an important amount of char is formed, confirming that graphite is an efficient intumescent agent as it can act as a carbonization compound.

However, it is important to highlight that in spite of achieving certain polymer ablation protection in this work due to the char formation in binary compositions, it was observed here a significant material weight loss (by *ca.* 65-75% - see Table 6) due to the considerable volatilization of degradation products, which possibly provoked the flame persistency up to the holding clamp, as well as the continuous flame dripping able to ignite the cotton (see Table 6).

The most interesting results were achieved for the ternary compositions, presenting the formation of a highly compact char on burning, able to significantly reduce the combustion rate as compared to binary compositions as well as preventing flame dripping and ignition of cotton (see Table 6), similar results were reported during horizontal burning tests of these specimens [17].

In the case of PLA-3CLO-3EG, the formation of a highly compact char was observed until completed sample burning. However, when the flame went up to the holding clamp the specimens were broken in two pieces provoking cotton ignition by the fallen polymer piece but not by flame dripping during combustion [17]. Contrarily, PLA-3CLO-(6-9)EG nanocomposites presented the formation of a very high compact char (see Table 6) which could considerably decreasethe material BR, as compared to PLA-3CLO-3EG and binary compositions, preventing the dripping of inflamed polymer and falling of sample pieces during combustion.

a) b)

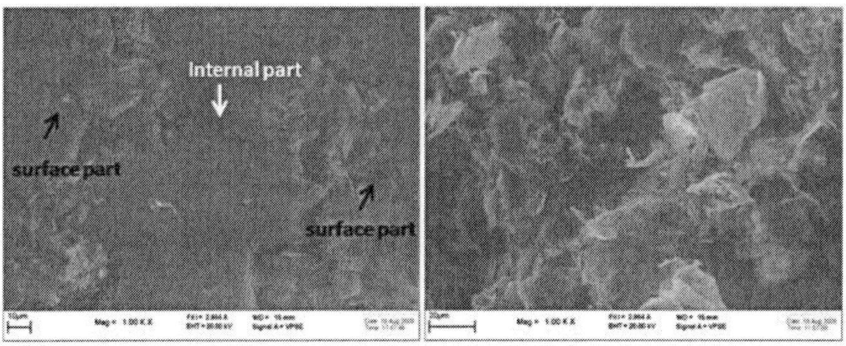

"Reprinted from Effect of expanded graphite/layered-silicate clay on thermal, mechanical and fire retardant properties of poly(lactic acid), 95/10,Kikku Fukushima, Marius Murariu, Giovanni Camino, Philippe Dubois, pages 1063-1074, Copyright (2010), with permission from Elsevier."

Figure 5. SEM micrographs of residual material from PLA-3EG after UL94 V test: (a) in the sample fracture zone and (b) on the surface part from (a).

Even more, for these PLA nanocomposites obtained by co-addition of CLO into PLA-EG compositions, the formation of the carbonized protective surface layer was confirmed through SEM analysis on the fracture surface of these materials (Figure 6). The surface of PLA-3CLO-3EG appeared as a continuous carbon layer rich in silicon, oxygen and expanded graphite layers (surface part in Figure 6a, Figure 6b), explained by the accumulation of both silicate and graphite layers on the sample surface. Similar to PLA-3CLO, the presence of high concentrations of Si (12%) was more noticeable in the surface as compared to the underlying material (indicated as "internal part" in Figure 6a, Si: 1.3%) [17].

A similar formation of surface carbon layer rich in silicate and graphite platelets was observed for the ternary compositions containing 6% or 9% EG (surface part in Figure 7a, Figure 7b), achieving for both materials a Si concentration by *ca.* 8 wt-% on the top layer whilst the smooth zones of the underlying material (indicated as "internal part" in Figure 7a) presented a Si concentration by *ca.* 1.1 wt-% comparable with pristine compositions. However, differently to what observed for PLA-3CLO-3EG residues, it was possible to observe for PLA-3CLO-(6-9)EG residues the continuous formation of a ceramic layer rich in both nanoparticles extending from the surface into the bulk material (circle in Figure 7a).

a) b)

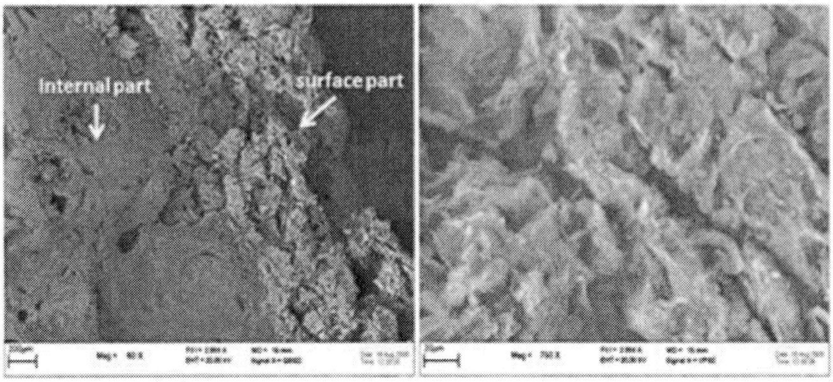

"Reprinted from Effect of expanded graphite/layered-silicate clay on thermal, mechanical and fire retardant properties of poly(lactic acid), 95/10,Kikku Fukushima, Marius Murariu, Giovanni Camino, Philippe Dubois, pages 1063-1074, Copyright (2010), with permission from Elsevier."

Figure 6. SEM micrographs of residual material from PLA-3CLO-3EG after UL94 V test (a) in the sample fracture zone and (b) surface part from (a).

a) b)

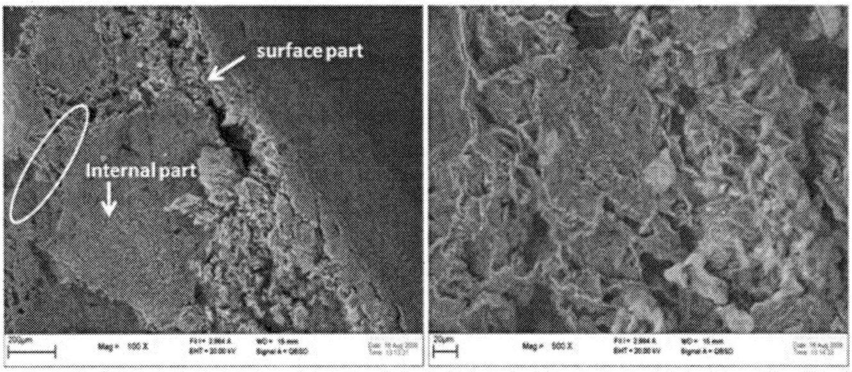

"Reprinted from Effect of expanded graphite/layered-silicate clay on thermal, mechanical and fire retardant properties of poly(lactic acid), 95/10,Kikku Fukushima, Marius Murariu, Giovanni Camino, Philippe Dubois, pages 1063-1074, Copyright (2010), with permission from Elsevier."

Figure 7. SEM micrographs of residual material from PLA-3CLO-9EG after UL94 V test (a) in the sample fracture zone and (b) surface part from (a). Similar SEM micrographs were obtained for PLA-3CLO-6EG.

Indeed, it was reported that this internal char presented a comparable Si concentration (*ca.* 7-8 wt-%) similar to that of the surface char [17]; and the formation of this extensive carbonaceous/ceramic layer, verified by means of ATR-FTIR and FT-IR analyses [17], was able to create a higher physical barrier towards progress of fire along the sample, polymer ablation and development of combustible volatile products as compared to PLA-3CLO-3EG and binary compositions, achieving the lowest BR for these compositions (see Table 6).

Moreover, the formation of this compact char layer rich in silicate and graphite platelets in ternary compositions was also found to considerably hinder mass transfer, obtaining lower material weight losses as compared to binary compositions (see Table 6) and indicating a considerable lower extent of underlying polymer degradation for these materials.

Finally, these investigations showed the remarkable flame retardant properties achieved for all compositions, especially for the ternary compositions due to the combined action of both CLO and EG components in the PLA matrix (particularly in nanocomposites with higher amount of expanded graphite [6-9]), regarding these last specimens as potentially interesting materials for the enhancement of PLA properties for technical applications. The deep understanding of the chemical interaction between these nanoparticles and the polymer could result in the design of poly(lactic acid) based nanocomposites with reduced combustion rate and self extinguishing behavior, thus contributing to the development of new fire-safe biobased polymer composites.

CONCLUSION

This chapter was addressed to review the effect of nanoparticles on the mechanical and fire retardant properties of a lactic acid based polymer -PLA-, which could present potential engineering applications.All PLA based nanocomposites showed considerable improvements in the mechanical and fire retardant properties of PLA. Due to the additive reinforcing effect of both nanofillers, the PLA-CLO-EG nanocomposites were characterized by considerably higher mechanical properties than those of PLA-EG nanocomposites.

The ternary systems revealed advanced flame retardant properties, attributed to formation of a ceramic/carbonaceous surface layer during burning. Even more, at high amount of graphite (6-9 wt-%) and by co-addition of clay (3 wt-%), the ternary compositions showed the formation of an extensive compact char, significant reduction of the vertical combustion rate and advanced anti-dripping properties.

ACKNOWLEDGMENTS

The author would like to thank greatfully Prof. Giovanni Camino, Prof. Philippe Dubois and Dr. Marius Murariu for their high support and guidance throughout this chapter. The author would like to acknowledge Miss Anne-Laure Dechief from LPCM for her very helpful collaboration and technical assistance, as well as COST action MP0701.

REFERENCES

[1] Södergård, A. and Stolt, M., *Progress in Polymer Science* 27, 1123 (2002)

[2] Ray, S. and Bousmina, M., *Progress in Materials Science* 50, 962 (2005).

[3] Hakkarainen, M., *Advances in Polymer Science* 157, 113 (2002).

[4] Chiellini, E., Chiellini, F. and Cinelli, P., *Degradable Polymers. Principles andApplications*, Luwer Academic Publishers, U.K. (2002).

[5] Dubois, Ph. and Murariu, M., *JEC Composites Magazine* 45, 66 (2008).

[6] Platt, D., *Biodegradable Polymers - Market report*, SmithersRapra Limited UK, Shawbury, Shrewsbury, Shropshire (2006)

[7] Solarski, S., Ferreira, M., Devaux, E., Fontaine, G., Bachelet, P., Bourbigot, S., Delobel, R., Coszach, P., Murariu, M., Da Silva Ferreira, A., Alexandre, M., Degee, Ph., andDubois, Ph., *J. Appl. Polym. Sci.* 109, 841 (2008).

[8] Pollet, E., Paul M.A. and Dubois, P., *Biodegradable Polymers and Plastics*, Kluwer Academic/Plenum Publishers, U.S. (2003).

[9] Utracki, L.A., *Clay- Containing Polymeric Nanocomposites*, Rapra Technology Limited, vol. 1, U.K. (2004).

[10] Pluta, M., Galeski, A., Alexandre, M., Paul M.A., and Dubois, P., *J. Appl. Polym.Sci.,* 86, 1497 (2002).

[11] Ogata, N., Jimenez, G., Kawai, H., and Ogihara, T., *J. Polym. Sci. Part B: Polym. Phys.* 35, 389 (1997).

[12] Bandyopadhyay, S., Chen, R. and Giannelis, E.P., *Polym. Mater. Sci. Eng.* 81, 159 (1999).

[13] Sinha Ray, S., Okamoto, K., Yamada, K., and Okamoto, M., *Nano Lett.* 2, 423 (2002).

[14] Sinha Ray, S., Yamada, K., Okamoto, M., and Ueda, K., *Nano Lett.* 2, 1093 (2002).

[15] SinhaRay, S., Maiti, P., Okamoto, M., Yamada, K., and Ueda, K., *Macromolecules*35, 3104 (2002).

[16] Fukushima, K., Tabuani, D. and Camino, G., *Materials Science and Engineering C* 29, 1433 (2009).

[17] Fukushima, K., Murariu, M., Camino, G., and Dubois, Ph., *Polymer Degradation and Stability* 95, 1063 (2010).

[18] Laoutid, F., Bonnaud, L., Alexandre, M., Lopez Cuesta, J.M., and Dubois, Ph., *Mater. Sci. Eng.R* 63, 100 (2009).

[19] Kalaitzidou, K., "Exfoliated graphite nanoplatelets as reinforcement for multifunctional polypropylene nanocomposites". PhD thesis, Michigan State University, Department of Chemical Engineering and Materials Science (2006).

[20] Hussain, F., Hojjati, M., Okamoto, M., and Gorga, R.E., *J. Compos. Mater.* 40, 1511 (2006).

[21] Yasmin, A., Luo, J.J. and Daniel, I.M., *Compos. Sci. Technol.* 66, 1182 (2006).

[22] Zhao, Y.F., Xiao, M., Wang, S.J., Ge, X.C., and Meng, Y.Z., *Compos. Sci. Technol.* 67, 2528 (2007).

[23] Kalaitzidou, K., Fukushima, H. and Drzal, L.T., *Compos. Sci. Technol.,* 67, 2045 (2007).

[24] Li, J., Kim, J.K. and Sham, M.L., *Scr. Mater.* 53, 235 (2005).

[25] Zheng, W., Wong, S.C. and Sue, H.J., *Polymer* 73, 6767 (2002).

[26] Huang, J.H., Baird, D.G. and McGrath, J.E., *J. Power Sources* 150, 110 (2005).

[27] Chen, X.M., Wei Shen, J. and Yi Huang, W., *J. Mater. Sci. Lett.* 21, 213 (2002).

[28] Chen, G., Weng, W., Wu, D., Wu, C., Lu, J., Wang, P., and Chen, X., *Carbon* 42, 753 (2004).

[29] Chen, G., Wu, D., Weng, W., and Wu, C., *Letters to the Editor / Carbon* 41, 579 (2003).

[30] Uhl, F.M., Yao, Q., Nakajima, H., Manias, E., and Wilkie, C.A., *Polym. Degrad. Stab.* 89, 70 (2005).

[31] Murariu, M., Dechief, A.L., Bonnaud, L., Paint, Y., Gallos, A., Fontaine, G., Bourbigot, S., and Dubois, Ph., *Polymer Degradation and Stability* 95, 889 (2010).

[32] Ciallella, C., Gruenberger, T.M., Grivei, E., and Probst, N., *Plastics Additives and Compounding* 10, 40 (2008).

[33] Solarski, S., Mahjoubi, F., Ferreira, M., Devaux, E., Bachelet, P., Bourbigot, S., Deloux, R.,Coszach, P., Murariu, M., Da Silva, A., Alexandre, M., Degee, P., and Dubois, P., *J. Mater. Sci.* 42, 5105 (2007).

[34] Paul, M.A., Alexandre, M., Degee, Ph., Henrist, C., Rulmont, A., and Dubois, Ph., *Polymer* 44, 443 (2003).

[35] Zanetti, M., Camino, G., Reichert, P., and Mülhaupt, R., *Macromol. Rapid Commun.* 22, 176 (2001).

[36] Chung, D.D.L., *J. Mater. Sci.* 37, 1475 (2002).

[37] Gilman, J., *Polymer-Clay (nano)composites,* Wiley and Sons Ltd, West Sussex, US (2000).

[38] Fina, A., Bocchini, S. and Camino, G., *Polym. Degrad. Stab.* 93, 1647 (2008).

In: Lactic Acid
Editors: A. Migallón and R. Ruseckaite
ISBN: 978-1-62081-643-1

Chapter 5

LACTIC ACID BACTERIA: AN APPROACH FOR NEW BIOPRESERVATION TECHNOLOGY

María R. Moreira,[1,2] Sara I. Roura,[1,2] Stella Echarte[2] and Alejandra G. Ponce[1,2]
[1]Consejo Nacional de Investigaciones Científicas y Técnicas (CONICET), Argentina
[2]Grupo de Investigación en Ingeniería en Alimentos, Facultad de Ingeniería, Universidad Nacional de Mar del Plata (UNMDP), Buenos Aires, Argentina

ABSTRACT

The industry of minimally processed vegetables is constantly searching for new technologies for maintaining quality and inhibiting undesired microbial growth in all steps of production and distribution. New biopreservation technologies are recently considered as a promising perspective in order to reduce the chemical preservatives. The aim of this study was to determine the influence of added bacteriocin-producing lactic acid bacteria (LAB) on native microbial population associated to lettuce. We also determined the "*in vivo*" protective effect of LAB against *E.coli O157:H7*. Lettuces were inoculated with single LAB strains (*Enterococcus faecium, Lactococcus lactis, Enterococcus hirae* and *Enterococcus canis*) or in combination with a pathogenic strain of *E.coli O157:H7*. The evolution of microorganisms in lettuce was followed during storage at 8 °C. Although the treatments did not completely eliminate the pathogen from lettuces, the results showed that LAB had a good colonization in both substrates. Biopreservation effect exerted by LAB on the *E.coli O157:H7* population was bacteriostatic for the case of *E.faecium* and *E.hirae.* Microbial competition strategies require time, which in products such as lettuce could produce the rapid quality deterioration. Therefore, it would be of great interest the study of these strains in products having a longer shelf life.

INTRODUCTION

Lactic acid bacteria (LAB) have long been used as starter cultures for foods and beverages fermentation for their contribution to flavor and aroma development and spoilage retardation. The preservative effect is not only due to acidic conditions that these bacteria create in foods, but also to their ability for producing and excreting inhibitory substances. These processes include hydrogen peroxide, ethanol, diacetyl, carbon dioxide, bacteriocin or antibiotic-like substances. Bacteriocins are defined as ribosomally synthesized antimicrobial peptides or proteins by LAB with antagonistic activity against species genetically close related [1]. Some of them are inhibitory towards food spoilage and foodborne pathogenic bacteria [2].

Numerous reviews have suggested that some LABs were able to control the growth of pathogens such as *Listeria monocytogenes* in food products [3,4]. LAB may compete for nutrients or space with spoilage microorganisms, such as certain Gram negative bacteria [5]. LAB cultures isolated from the same type of vegetables or products in which they will be subsequently used as biocontrol agents may have the greatest success chance in controlling pathogens [6,7]. The exploration of naturally occurring antimicrobials in food preservation receives increasing attention due to consumer awareness of natural food products and a growing concern about microbial resistance toward conventional preservatives. Organic vegetables are becoming increasingly popular among modern consumers. Ponce *et al.* [8] characterizing the native microbial population on Swiss chard cultivated by organic methods reported that to achieve the same level of removal in organic leaves, harsher washing procedures are needed compared with leaves obtained by traditional methods. During the post-harvest handling of organic vegetable it precludes the use of chemical additives. For this reason, harmless LAB exhibiting antagonistic effects against pathogenic bacteria may be of applied interest as bio-preservatives of minimally processed organic vegetables. Several studies addressing the bacteriocins produced by LAB isolated from dairy, meat and fish products have been reported in the literature [9,10] but studies on bacteriocin-producing LABs isolated from organic vegetables are very scarce.

On the other hand, the increasing popularity of salad bars offering freshly cut lettuces, which are also found in supermarkets and convenience stores, has introduced new environments which support the growth of food-borne pathogens including *Escherichia coli.* Most of these products are perishable and require protection from spoilage during preparation, storage, and distribution to obtain a useful shelf life [11]. Ready-to-use vegetable salads generally contain few or no antimicrobial additives, and rely on refrigeration as the main preservation factor. The kind of processing to which the ready to use products are subjected makes them particularly sensitive to microbial contamination. A variety of pathogens have been found on fruit and vegetable products, including *Salmonella* and *Shigella spp.*, enteropathogenic strains of *E. coli*, *Aeromonas hydrophila*, *Yersinia enterocolitica*, and *Staphylococcus aureus* [12]. These pathogens grow and cause diseases depending on the type of product, storage conditions (time, temperature, and atmosphere) and competitive microflora. Thus, there is a particular interest to introduce additional safety measures for these products. Efficient inhibition of the pathogens growth by antibacterial-producing lactic acid bacteria (LAB) on ready-to-use vegetables may be achieved by strains that grow and secrete these compounds under refrigeration conditions [13,14]. Lactic acid bacteria compete with other microbes by modifying the microenvironment by their metabolic products [13-15]. The ecological adaptation of added LAB is an important factor for their effectiveness as biocontrol cultures, and their isolation from the same or similar vegetables on which they will be further used as inhibitory agent could be interesting.

The goal we will describe in this chapter was to investigate LAB with antibacterial activity within the facility microbial organic ecosystem and finally to select interesting strains for the role of biopreservatives. Then we will determine the protective ability and colonization of four LAB strains producing bacteriocin like substances (*Enterococcus faecium, Lactococcus lactis, Enterococcus hirae and Enterococcus canis*) applied to lettuce and the LAB effect exerted on *E. coli* growth.

SCREENING FOR LAB ANTAGONISTIC ACTIVITY

LABs isolated from vegetables are probably the best candidates for improving the microbiological safety of fresh foods, because they are well adapted to the conditions in these kinds of vegetables and should therefore be more competitive than LABs from other sources. The reiterative use of synthetic chemicals in the conventional methods of cultivation could be interfering with the anchorage of the native microflora making it easier to remove. In organic vegetables, the microflora would have more time and gentler conditions to proliferate, making bonds stronger with the substrate [8]. The well known agar diffusion assay was used to study the antimicrobial activity of the 45 LAB strains isolated from organic Swiss chard and spinachs, intercrossing the strains between them as indicators. Four strains were shown to produce inhibition zones against indicators and were identified by ribosomal DNA genes analysis as *Lactococcus lactis* subsp *lactis* and *Enterococcus canis* (isolated from spinachs), *Enterococcus hirae* and *Enterococcus faecium* (isolated from Swiss chard). Inhibitory spectra of these isolates are presented in Tables 1 and 2. The antimicrobial activity of LABs may be due to the production of organic acids, with a consequent reduction in pH or to the production of hydrogen peroxide. It may also be due to the production of bacteriocins or bacteriocins-like compounds [16]. Hence antimicrobial activity was checked after excluding inhibition due to organic acids and hydrogen peroxide. The supernatant from *L.lactis* showed the broadest spectrum by inhibiting all the microorganism indicators, while the supernatant from *E.faecium* presented antagonistic effects against *L.plantarum* and *L.monocytogenes*. On the other hand, *E.hirae* supernatant was less effective against *L.monocytogenes* than *E.faecium*. In agreement with these results, others authors reported numerous Enterococci strains associated with food systems, mainly *E.faecalis* and *E.faecium* capable of producing a variety of bacteriocins, called enterocins, with activity against *L.monocytogenes* [17-19]. *E.canis, L.lactis* and *E.hirae* were able to inhibit GIIA 30, a taxonomically related microorganism isolated from the same raw material. Wild bacterial strains need to withstand the competition of other microorganisms to survive in their hostile natural environment, so they often produce antimicrobials [20]. This fact could explain the antimicrobial activity that the selected strains presented for taxonomically related microorganisms derived from the same raw material (Table 1). An inhibitory spectrum of selected LAB to inhibit Gram-negative foodborne pathogens was also tested. The capacity of LABs to inhibit Gram-negative microorganisms is presented in Table 2. Two of the four

E.coli strains utilized as indicators were inhibited by *E.faecium* and *L.lactis*, while three of the pathogen strains were inhibited by *E.hirae* and *E.canis*. The supernatants of *E.faecium*, *E.hirae* and *E.canis* were slightly capable to inhibit *A.xylosoxidan*. Such bacteriocin-like inhibition of a Gram negative bacteria by a Gram-positive bacteria is still relatively novel. Few works have reported similar facts, *L.lactis* and *Pediococcus acidilactici* isolated from alfalfa sprouts showed competitive inhibition against *E.coli O157:H7* [21]. The bacteriocin-like substance produced by *Lactobacillus plantarum TF711* from Tenerife cheese was active against Gram-negative bacteria, such as *Shigella sonnei* and *Klebsiella pneumoniae* [22]. Talarico *et al.* [23] previously found a broad spectrum of substances from *Lactobacillus reuteri* with antimicrobial activity against Gram-negative bacteria.

The 4 strains used to enrich the microflora of lettuce leaves were isolated and characterized from plant substrates by Ponce *et al.* [24] as mentioned above. These strains have been shown to present *in vitro* inhibitory activity on Gram-negative bacteria (Table 1 and 2). Figure 1 shows the results obtained from inoculated lettuce with the 4 strains of lactic acid bacteria producing metabolites simil bacteriocins; *Enterococcus faecium, Lactococcus lactis, Enterococcus hirae* and *Enterococcus canis*. For each treated and control samples, we performed total microflora counts, lactic acid bacteria and endogenous *E.coli* during storage.

Figure 1A shows the evolution of the total microflora in each samples of lettuce over 120 hours of refrigerated storage. The evolution of the total microflora showed no significant differences ($p < 0.01$) between samples along the storage time.

Figure 1B indicates that the evolution of lactic acid bacteria during storage showed no significant differences between samples throughout the storage. This could be explained by 3 possible hypothesis: first, the endogenous lactic acid bacteria present in lettuce prevent the colonization of lactic acid bacteria inoculated exogenously. Secondly, the opposite occurs, LAB inoculated successfully colonize the substrate and prevent the growth of endogenous lactic acid bacteria population. Finally, it could happen that the storage time was not sufficient to deploy exogenous LAB good competitive strategies of substrate colonization. However, endogenous or inoculated lactic acid bacteria in lettuce leaves had a significant increase during storage, indicating that lettuce is a good substrate for this type of bacteria [24,25].

Figure 1C showed endogenous *E.coli* counts in control samples and those inoculated with LAB during storage. The control samples showed a linear increase in counts of endogenous *E.coli* with initial and final values of 4.71

and 6.72 log CFU g^{-1} respectively. The *E.coli* counts for samples inoculated with *E.faecium* and *E.hirae* remained constant in both samples during storage, with approximate initial and final values of 5.20 and 5.40 log CFU g^{-1} respectively, for both samples. In the case of samples that were supplemented with *E.canis* showed a similar pattern with constant values of *E.coli* to 96 hours of storage. These results could be indicating a bacteriostatic effect of these three strains of LAB on the growth of endogenous *E.coli.* Samples enriched with *L.lactis* and *E.canis* indicated an increase in the counts (Figure 1C) compared to control, which could indicate a lack of biocontrol of *E.coli* endogenous at 96 hours by *L.lactis* and 120 hours by *E.canis*.

Table 1. Inhibitory effects of LAB supernatants on Gram-positive indicator microorganisms

Producer strain	Indicator inhibition strains							
	L. bulgaricus			*S. thermophilus*	*GIIA*	*L. plantarum*	*P. pentasaceus*	*L. monocytogenes*
	LB1	LB2	LB3	ST1	30	CRL691	ATCC10791	
E.faecium	-[a]	-	-	-	-	12.7± 2.1	-	14.4±0.9
L. lactis	8.0±1.0	10.0±1.0	10.0±1.0	-	10.3 ±2.1	9.2±3.3	11.0±3.3	6.5±0.5
E. hirae	-	-	-	-	9.6 ±1.5	-	-	7.1±0.9
E. canis	-	-	-	-	11.0 ±1.0	-	-	-

LB1, LB2 and LB3: *Lactobacillus delbrueckii* subsp. *bulgaricus* isolated from dairy products. ST1: *Streptococcus thermophilus* isolated from dairy products. GIIA: Lactic Acid Bacteria isolated from Grupo de Investigación en Ingeniería en Alimentos (Buenos Aires, Argentina); CRL 691: *Lactobacillus plantarum* and ATCC10791: *Pedicoccus pentasaceus*.

-[a]: inhibition zones were expressed in mm. Halos ≤ 6 mm were considered negative inhibition.

To eliminate the inhibitory effects produced by organic acids the supernatants were adjusted to pH 6.5. The hydrogen peroxide effect was discarded using catalase.

Values presented are the mean of five independents experiments by duplicate ± SD.

Table 2. Inhibitory effects of LAB supernatants on Gram-negative indicator microorganisms

Producer strain	Indicator strains				
	E. coli CI	*E. coli CII*	*E. coli 25922*	*E. coli 32158*	*A. xylosoxidans*
E. faecium	10.5±0.6	-	10.1±2.0	-	6.0±0.0
L. lactis	11.0±1.34	-	-	12.0±1	-
E. hirae	10.9±0.9	-	9.1±1.9	10.6±0.9	7.0±1.0
E. canis	-[a]	11.3±1.5	10.8±1.5	13.3±1.5	6.0±0.0

CI and CII: *Escherichia coli* isolated from meat, ATCC 25922 and ATCC 32158: *Escherichia coli* from American Type Culture Collection.

-[a]: inhibition zones were expressed in mm. Halos ≤ 6 mm were considered negative inhibition.

To eliminate the inhibitory effects produced by organic acids the supernatants were adjusted to pH 6.5. The hydrogen peroxide effect was discarded using catalase.

Values presented are the mean of five independents experiments by duplicate ± SD.

ENRICHMENT OF LETTUCE LEAVES BY CO-INOCULATION WITH LACTIC ACID BACTERIA STRAINS PLUS *E. COLI O157: H7*

Figure 2 (A, B and C) shows the results of total microflora, lactic acid bacteria and *E.coli* endogenous resulting in lettuce samples co-inoculated with LAB strains and *E.coli O157: H7*. In this case, lettuce samples inoculated only with *E.coli O157: H7* were taken as control. The evolution of the total microflora in all treated samples showed no significant differences ($p>0.01$) between them. In all cases, there was an increase in microbial counts of 3.54 orders log CFU g^{-1} approximately (Figure 2A). Similar results were obtained for lactic acid bacteria counts in all samples, although they increased their counts during storage at about 3.35 orders log CFU g^{-1}, without significant differences between them (Figure 2B). Figure 2C shows that *E.coli* counts in samples inoculated with *E.coli O157: H7* and co-inoculated with LAB strains (*E.faecium, L.lactis, E.hirae* and *E.canis*) showed significant differences ($p <0.01$) after 48 hours of storage.

a)

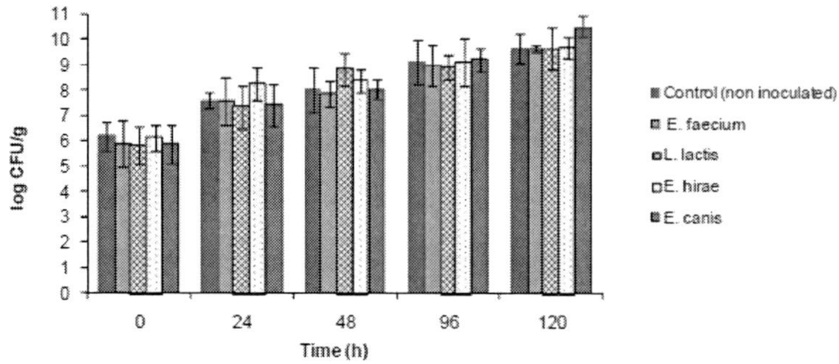

b)

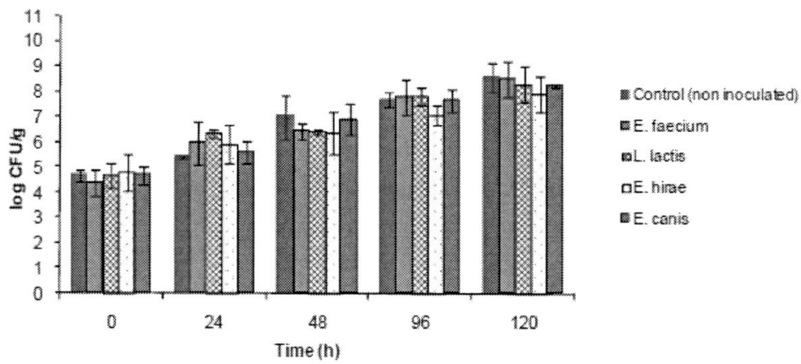

c)

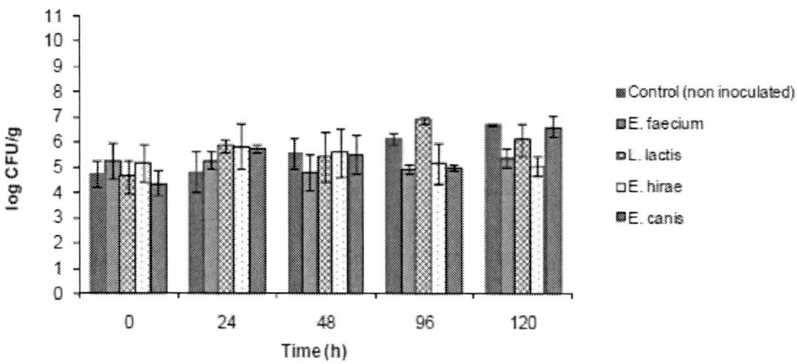

Error bars denote standard deviation from the mean values (n=4).

Figure 1. Microbial counts in lettuce inoculated with selected LAB strains and storage at 8°C. Total microflora counts (A); lactic acid bacteria (B) and endogenous E. coli (C).

a)

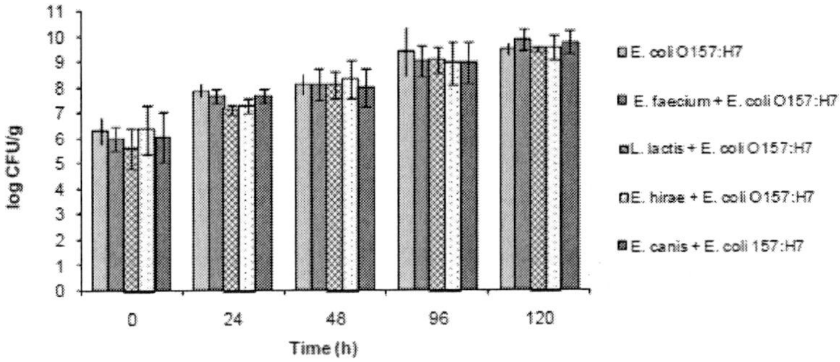

b)

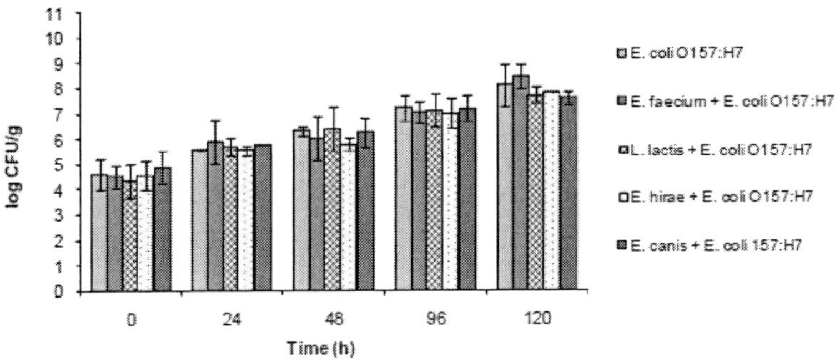

c)

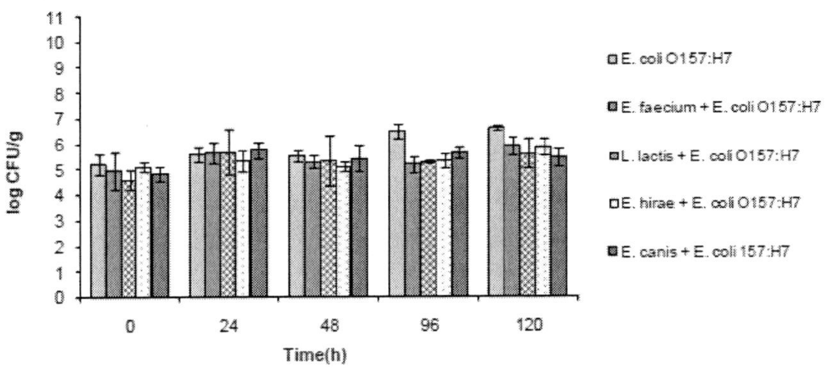

Error bars denote standard deviation from the mean values (n=4).

Figure 2. Microbial counts in lettuce co-inoculated with selected LAB strains and E.coli O157:H7, storage at 8°C. Total microflora counts (A); lactic acid bacteria (B) and endogenous E.coli (C).

In these samples LAB had a bacteriostatic behavior on endogenous *E.coli* growth during the storage, except for the sample co-inoculated with the pathogen and *L.lactis* where this effect begins within 24 hours and continued until the end of storage, up to 120 hours.

This bacteriostatic effect observed in the samples co-inoculated with LAB could be due to a competitive advantage they had against *E.coli O157: H7* to colonize the substrate or the release of metabolites simil bacteriocins from inoculated LAB. Wilderdyke *et al.* [21] found that lactic acid bacteria strains isolated from alfalfa sprouts showed competitive inhibition against *E. coli*. As was mentioned above, Ponce *et al.* [24] in *in vitro* studies found a clear inhibition of *E.faecium* and *L.lactis* against 4 strains of *E.coli* used as indicators, while 3 strains of *E.coli* pathogens were inhibited by *E.canis* and *E.hirae*. In a previous research, has been shown that most of the bacteriocins were associated to bacterial growth and the effectiveness of these bacteriocins is affected by factors such as the type of strain, growth medium composition and environmental conditions of fermentation [26]. The effectiveness of "simil bacteriocins" metabolites depends on the physical and chemical properties of food used as a substrate. Furthermore, Allende *et al.* [25] analyzed the growth and production in a model system of food such as lettuce (lettuce extract). Authors also found that one of the factors affecting the production of antimicrobial substances by LAB were the refrigerated temperature storage. These authors found that for temperatures such as 4 and 10 °C only nisin Z-producing strain (IPLA 1064) was able to produce bacteriocin, although all strains tested showed good growth at temperatures above. On the other hand, these same authors showed that at 32 °C the production of bacteriocins by the above strains was effective except for the production of plantaricin C.

CONCLUSION

LABs isolated from the same vegetables on which they will be further used as biocontrol cultures directly added to vegetables ensure that these LAB strains are well ecological adapted. This fact is an important factor for their effectiveness as natural antimicrobial agents. The native LAB isolated from organic spinach and Swiss chard presented antagonistic activity against frequently foodborne pathogens, such as *L.monocytogenes* and *E.coli*.

The population of endogenous *E.coli* showed advantage to colonize the substrate respect to *E.coli O157:H7*. In agreement with these results, the same effect was observed when samples were enriched with each lactic acid bacteria, since counts for this population did not differ respect to samples without enrichment. Possibly, the inocula used to enrich the microflora of lettuce itself require a longer time of colonization due to energy expenditure needed to compete against the new substrate. Biopreservative effect exerted by LAB on the populations of *E.coli* was bacteriostatic for the case of *E.faecium* and *E.hirae*, as the pathogen counts remained constant throughout storage. These strains would exert as a biopreservative and could be used in combination with another type of obstacle. Investigate the importance of cover crops such as LAB, to exert a biocontrol would aim to balance the dynamics of microbial populations, favoring a bacteriostatic control bactericidal against those methods in order to avoid development of resistant pathogenic strains.

In general, it is required more knowledge about the mechanisms involved in the physiological response of microorganisms, against the action of natural preservation technologies applied both individually and combined, and further study of food and its environment, to implement technology barriers that are most effective combination, such as modified atmosphere or other environmentally-friendly techniques.

REFERENCES

[1] De Vuyst, L. and Vandamme, E. (1994). Bacteriocins of lactic acid bacteria: Antimicrobial potential of lactic acid bacteria. London: Blackie In: L. De Vuyst, and E. Vandamme (Eds.), (pp. 91–142).

[2] Zamfir, M., Callewaert, R., Cornea, P., Savu, L., Vatafu, I., and De Vuyst, L. (1999). Purification and characterization of bacteriocin produced by Lactobacillus acidophilus IBB 801. *Journal of Applied Microbiology*, 87, 923–931.

[3] Callewaert, R., Hugas, M. and De Vuyst, L. (2000). Competitiveness and bacteriocin production of Enterococci in the production of Spanishstyle dry fermented sausages. *International Journal of Food Microbiology*, 5, 33–42.

[4] Mataragas, M., Metaxopoulos, J., Galiotou, M., and Drosinos, E. (2003). Influence of pH and temperature on growth and bacteriocin production by *Leuconostoc mesenteroides* L124 and *Lactobacillus curvatus* L442. *Meat Science*, 64, 265–271.

[5] Campos, A., Rodríguez, O., Calo-Mata, P., Prado, M., and Barros-Velázquez, J. (2006). Preliminary characterization of bacteriocins from *Lactococcus lactis*, *Enterococcus faecium* and *Enterococcus mundtii*

strains isolated from turbot (Psetta maxima). *Food Research International*, 39, 356–364.

[6] Vescovo, M., Torriani, S., Orsi, C., Macchiarolo, F., and Scolari, G. (1996). Application of antimicrobial-producing lactic acid bacteria to control pathogens in ready-to use vegetables. *Journal of Applied Bacteriology*, 81, 113–119.

[7] Breidt, F. and Fleming, H. (1997). Using lactic acid bacteria to improve the safety of minimally processed fruits and vegetables. *Food Technology*, 51, 46–51.

[8] Ponce, A., Fritz, R., Del Valle, C., and Roura, S. (2003). Antimicrobial activity of essential oils on native microbial population of organic Swiss chard. *Lebensmittel-Wissenschaft und-Technology*, 36, 679–684.

[9] Marrug, J. (1991). Bacteriocins, their role in developing natural products. *Food Biotechnology*, 5, 305–312.

[10] Campos, A., Rodríguez, O., Calo-Mata, P., Prado, M., and Barros-Velázquez, J. (2006). Preliminary characterization of bacteriocins from *Lactococcus lactis*, *Enterococcus faecium* and *Enterococcus mundtii* strains isolated from turbot (Psetta maxima). *Food Research International*, 39, 356–364.

[11] Holley, R. and Patel, D. 2005. Improvement in shelf-life and safety of perishable foods by plant essential oils and smoke antimicrobials. *Food Microbiol.* 22, 273–292.

[12] Nguyen-THE, C. and Carlin, F. 1994. The microbiology of minimally processed fresh fruits and vegetables. Crit. Review Food Sci. *Nutrit.* 34, 371-401.

[13] Gomez, D., Miliwebsky, E., Fernandez Pascua, C., Baschkier, A., Manfredi, E., Zotta, M., Nario, F., Piquín, A., Sanz, M., Etcheverría, A., Padola, N., Parma, A., and Rivas, M. (2002). Aislamiento y caracterización de *Escherichia coli* productor de toxina Shiga en hamburguesas supercongeladas y quesos de pasta blanda. *R. Argentina Microbiología* 34, 66–71.

[14] Vescovo, M., Gianluigi, S. and Zacconi, C. (2006). Inhibition of *Listeria innocua* growth by antimicrobial-producing lactic acid cultures in vacuum-packed cold-smoked salmon. *Food Microbiol.* 23, 689–693

[15] Niku-Paavola, M, Laitila, A., Mattila-Sandholm, T., and Haikara, A. (1999). New types of antimicrobial compounds produced by *Lactobacillus plantarum*. *J. Appl. Microbiol.* 86, 29–35

[16] González, L., Sandoval, H., Sacrista´ n, N., Castro, J., Fresno, J., and Tornadijo, M. (2007). Identification of lactic acid bacteria isolated from

Genestoso cheese throughout ripening and study of their antimicrobial activity. *Food Control*, 18, 716–722.

[17] Herranz, C., Casaus, P., Mukhopadhyay, S., Martínez, J., Rodríguez, J., and Nes, I. (2001). *Enterococcus faecium* P21: A strain occurring naturally in dry-fermented sausages producing the class II bacteriocins enterocin A and enterocin B. *Food Microbiology*, 18(2), 115–131.

[18] Sarantinoupoulos, P., Kalantzopoulos, G. and Tsakalidou, E. (2002). Effect of Enterococcus faecium on microbiological, physicochemical and sensory characteristics of Greek Feta cheese. *International Journal of Food Microbiology*, 76, 93–105.

[19] Giraffa, G. (2003). Functionality of enterococci in dairy products. *International Journal of Food Microbiology*, 88, 215–222.

[20] Ayad, E., Verheul, A., Jing, C., Wouters, J., and Smit, G. (2002). Antimicrobial producing wild lactococci from artisanal and non-dairy origins. *International Dairy Journal*, 12, 145–150.

[21] Wilderdyke, M., Smith, D. and Brashears, M. (2004). Isolation, identification and selection of lactic acid bacteria from alfalfa sprouts for competitive inhibition of foodborne pathogens. *Journal of Food Protection*, 67(5), 947–951

[22] Hernández, D., Cardell, E. and Zárate, V. (2005). Antimicrobial activity of lactic acid bacteria isolated from Tenerife Cheese: initial characterization of plantaricin TF711, a bacteriocin-like susbstance produced by Lactobacillus plantarum TF711. *Journal of Applied Microbiology*, 99(1), 77–88.

[23] Talarico, T., Casas, I., Chung, T., and Dobrogosz, W. (1988). Production and isolation of reuterin, a growth inhibitor produced by Lactobacillus reuteri. *Antibacterial Agents Chemotherapy*, 32, 1854–1858.

[24] Ponce, A., Moreira, M., Del Valle C., and Roura, S. (2008). Preliminary characterization of bacteriocin-like substances from lactic acid bacteria isolated from organic leafy vegetables. *LWT* 41, 432-41.

[25] Allende, A., Martínez, B., Selma, V., Gil, M., Suárez, J., and Rodríguez, A. (2007). Growth and bacteriocin production by lactic acid bacteria in vegetable broth and their effectiveness at reducing *Listeria monocytogenes* in vitro and in fresh-cut lettuce. *Food Microbiol.* 4, 759-766.

[26] Parente, E. and Ricciardi, A. (1999). Production, recovery and purification of bacteriocins from lactic acid bacteria. *Applied Microbiology Biotechnology*, 52, 628–638.

In: Lactic Acid
Editors: A. Migallón and R. Ruseckaite

ISBN: 978-1-62081-643-1

Chapter 6

PRODUCTION OF LACTIC ACID THROUGH AGROINDUSTRIAL WASTES. THE CASE OF CHEESE WHEY EXPLOITATION

S. Plessas,[1,*] ***A. Alexopoulos***[1] ***and E. Bezirtzoglou***[1,2]

[1]Democritus University of Thrace, Faculty of Agricultural Development, Department of Food Science and Technology, Laboratory of Microbiology, Biotechnology and Hygiene, Greece

[2]Democritus University of Thrace, Faculty of Agricultural Development, Department of Food Science and Technology Laboratory of Food Processing, Greece

ABSTRACT

Lactic acid is a very important chemical compound for food, chemical and pharmaceutical industries. Regarding the food industry applications of lactic acid, it is generally used as natural preservative and as an acidulant. It is usually produced through fermentation of various raw materials. However, it is estimated that approximately 90% of the total production cost of lactic acid is due to these raw materials. Likewise, the research community makes efforts to develop new technologies and to exploit attractive agro-industrial by-products targeting to the decrease of the whole production cost. In this pattern, corn sugars, molasses and

* Corresponding author: Dr Plessas Stavros, GR68200, Greece. Tel and fax: +30 2552041141. E-mail address: splessas@agro.duth.gr.

cheese whey have been extensively examined. However, cheese whey seems to offer the more preferred raw material, since it is cheap, abundant and responsible for environmental pollution due to its high organic load. Several studies have been published the last decade, suggesting ways for the exploitation of cheese whey. On the other hand, the appropriate choice of the bacteria that will degrade cheese whey is of great importance. For this purpose several lactic and bacteria (LAB) have been applied and quite encouraging results have been emerged. Among LAB *Lactobacillus helveticus* is the most effective, due to the achievement of higher lactic acid productivities compared to other LAB. In addition application of *Streptococcus* has also led to satisfying results. Finally, genetic modification for the development of a satisfactory and efficient LAB strains for cheese whey fermentation conduction has also been studied, as well as various immobilization techniques that have also been successfully applied. In the frame of this review the importance of lactic acid and the most significant research efforts for enhancing lactic acid production are discussed.

INTRODUCTION

Lactic acid is probably one of the most important chemical compound for food, chemical, textile and pharmaceutical industries. Lactic acid is a carboxylic acid with the chemical formula $C_3H_6O_3$. Specifically it has a hydroxyl group adjacent to the carboxyl group, making it an alpha hydroxy acid. It was first isolated by Carl Wilhelm Scheele in 1780 as the acid of milk and produced commercially by the German pharmacy Boehringer Ingelheim in 1895 (Gregor 1999). However, the first commercially manufactured form of calcium lactate was conducted in the US in 1883 at the Avery Lactate Company, Littleton, Massachusetts. Lactic acid can be produced either by chemical synthesis or fermentation. The main procedure regarding the chemical synthesis of lactic acid is the hydrolysis of lactonitrile by strong acids, which provide a racemic mixture of D- and L-lactic acid. (Litchfield, 1996; Datta, 1995). It is estimated that approximately 10% of lactic acid is produced through chemical synthesis annually. The rest (90%) is produced through microbial fermentation. The main reason is that through the microbial fermentation an optically pure product can be obtained, whereas synthetic production always results in a racemic mixture of D and L- lactic acid. In 2008, global production of lactic acid was estimated over 500,000 tonnes with current production between 300,000-400,000 tonnes per year. (NNFCC Renewable Chemicals Factsheet: Lactic Acid).

There are many applications of lactic acid for different purposes. Specifically, lactic acid is widely used as natural preservative and as an acidulant in the food and beverage sector. Besides this application the chemical industry uses lactic acid as solvent and material in the production of lactate ester. Through polymerization of lactic acid, a biodegradable polymer polylactic acid (PLA) is produced. Polylactic acid (PLA) polymer is currently one of the most promising biopolymers with many applications. For example PLA is widely used in medical applications as surgical implants (Jain, 2000), wound closures and green solvents (Litchfield, 1996). PLA is also used in food packaging, with remarkable results. Specifically, an antimicrobial packaging system based on PLA seems more efficient compared to other antimicrobial systems due to its low cost, effective antimicrobial activity, and environmental friendly technology. In the frame of this scientific survey the most important parameters for microbial fermentation of agroindustrial wastes for lactic acid production in food industries will be discussed.

Microbial Fermentation for Lactic Acid Production

As it was previously mentioned, the main advantage of microbial production of lactic acid compared to chemical synthesis is that in the first case lactic acid is produced as an optically pure product. However, the feasibility and amelioration of the microbial type of lactic acid production also depends on other parameters. The major parameters are the application of the suitable microorganism in a free or immobilized type that will be applied for the fermentation, the fermentation process that will be conducted and the proper selection of the agro-industrial waste that will be fermented. The targets for an efficient microbial lactic acid production are the achievement of high concentrations of produced lactic acid, high lactic acid yields and high lactic acid productivities. Regarding the lactic acid yield, the theoretical value is estimated at 71% (Tango and Ghaly, 1999). Thus the design of a fermentation system for lactic acid production needs the evaluation of all the following critical parameters, in order to achieve the best results. It is rational to target to high levels for all the parameters discussed. However there are some priorities that are obliged in some cases. For example when the raw material is not cheap then the main target is to maximize lactic acid yield. The same are forced in the case of semicontinuous operation, whereas the volumetric productivity is maximized by continuous operation if investment costs are high (Hofvendahl, Hahn–Hagerdal, 1997).

Microorganisms Applied for Lactic Acid Production

Numerous microorganisms have been applied for lactic acid production through the fermentation of various pure or agro-industrial substrates. In general, fungi and bacteria are the predominant species for lactic acid production through fermentation of various substrates. In particular, filamentous fungi gain more attention the last years for the production of lactic acid. *Rhizopus spp.*, have advantages compared to the bacteria, including their amylolytic characteristics and low nutrient requirements. However, fungal fermentation leads to the formation of by-products, such as ethanol, reducing the lactic acid yield. Some attempts regarding fungi applications for lactic acid production are recapitulated to the following: (i) *Rhizopus oryzae* was applied for the fermentation of potato waste water and waste paper (Vijayakumar, Aravindan and Viruthagiri, 2008); (ii) *Rhizopus oryzae* AS 3.819 was applied for lactic acid production in a semicontinuous fermentation achieving high L-lactic acid yield and volumetric productivity (Wu, Jiang, Liu, Pan, Zheng, Luo, 2011); (iii) *Rhizopus arrhizus* 36017 and *Rhizopus oryzae* 2062 were studied in the simultaneous saccharification and fermentation (SSF) of potato starch wastewater for lactic acid production. *Rhizopus arrhizus* 36017 showed a higher ability to produce lactic acid, while *Rhizopus oryzae* 206 produced more fungal biomass under similar conditions (Huang, Jin, Lant and Zhou, 2005). On the other hand lactic acid bacteria (LAB) are considered as the predominant species for lactic acid production.

Lactic acid bacteria are Gram-positive, non-sporing and are categorized as facultative anaerobes. Lactic acid bacteria have limited biosynthetic capabilities, thus they require many vitamins (especially B vitamins) amino acids, purines, and pyrimidines (Prescott 1996). They are classified into 3 groups based on their metabolism of sugars: Group 1, species that are homofermenters of hexoses to lactate and do not ferment pentoses. Group 2, species that are facultative heterofermenters of hexoses which means that they ferment hexoses by glycolysis to lactate about 50% or, under glucose limitation, to lactate, acetate, ethanol, and formate. Also, they ferment pentoses to lactate and acetate by the phosphoketolase pathway. Finally Group 3 includes species that are heterofermenters of sugars.

In general, *Lactobacillus delbrueckii* emerges as the one of the most efficient strains for lactic acid production, as it results to high lactic acid concentrations, yields and sometimes productivities (Hofvendahl, Hahn–Hagerdal, 2000). It is also noteworthy that *Lactobacillus delbrueckii* has been subjected to mutagenesis for the enhancement of its tolerance to lactic acid. The mutants resulted in improved lactic acid production than the wild type, proving the efficiency of this new strain.

There are also and other strains that are considered as efficient strains for lactic acid production, such as *Streptococcus thermophilus, Lactobacillus acidophilus, Lactobacillus rhamnosus, Lactococcus. lactis* (Table 1).

Table 1. Comparison of different strains and substrates for lactic acid production

Organism	Substrate	Lactic acid concen.	Lactic acid yield	Lactic acid product.	Reference
		(g L^{-1})	(g/g)	(g L^{-1} h^{-1})	
Lactobacillus casei NRRL B-441	glucose	82	0.91	5.6	Hujanen and Linko, 1996
Lactobacillus rhamnosus ATCC 10863	glucose	68	0.76	3.5	Hujanen and Linko, 1996
Lactobacillus delbrueckii mutant DP3 19	glucose	68	0.57	1.2	Demirchi and Pometto, 1992
Lactobacillus amylovorus ATCC 33622	hydrolyzed barley flour	93	0.52	2.0	Javanainen and Linko, 1995
Lactobacillus casei NRRL B-441	hydrolyzed barley flour	120	0.67	1.5	Javanainen and Linko, 1995
Lactobacillus delbrueckii sp. bulgaricus ATCC 11842	sorghum	-	-	4.5	Samuel and Lee,1980

Table 1. (Continued)

Organism	Substrate	Lactic acid concen.	Lactic acid yield	Lactic acid product.	Reference
		(g L-1)	(g/g)	(g L-1 h-1)	
Lactobacillus plantarum ATCC 14917	sorghum	-	-	2.0	Samuel and Lee,1980
Lactobacillus rhamnosus CCM 1753	lignocellulose hydrolyzed	37	0.74	-	Melzoch, Votruba, Habova, Rychtera, 1997
Lactobacillus delbrueckii sp. bulgaricus AU	molasses	20	0.45	-	Tiwari, Pandey, Mishra, 1979
Lactobacillus delbrueckii sp. delbrueckii ATCC 9649	molasses	26	0.58	-	Tiwari, Pandey, Mishra, 1979
Lactobacillus rhamnosus ATCC 7469	molasses	18	0.40	-	Tiwari, Pandey, Mishra, 1979
Lactococcus. lactis sp. lactis ATCC 19435	hydrolyzed wheat flour	96	0.76	3.0	Hofvendahl, 1997
Lc. lactis sp. lactis AS211	hydrolyzed wheat flour	95	0.77	1.7	Hofvendahl, 1997
Enterococcus faecium	hydrolyzed cod+ corn syrup	11	0.45	0.10	Giurca and Levin, 1993
Lactobacillus plantarum	hydrolyzed cod+ corn syrup	17	0.70	0.10	Giurca and Levin, 1993
Lactobacillus sp. strain FCP2	sugar-cane juice	104	2,8	96	Timbuntam, Sriroth, Tokiwa, 2006
Lactobacillus bulgaricus ACCC 11058	Food waste	35.12		0.98	Wang, Xu, Qiu, Sun, Han, Ming Wang, 2010
Lactobacillus	sugarcane	104	0.90	1.48	Buenaventurad

Organism	Substrate	Lactic acid concen.	Lactic acid yield	Lactic acid product.	Reference
		(g L-1)	(g/g)	(g L-1 h-1)	
delbrueckii JCM 1148	molasses				a, Calabi, Tokiwa, 2007
Lactobacillus delbrueckii JCM 1148	sugarcane juice	118	0.95	1.16	Buenaventurada, Calabi, Tokiwa, 2007
Lactobacillus delbrueckii JCM 1148	sugar beet juice	82	0.88	1.66	Buenaventurada, Calabi, Tokiwa, 2007

FERMENTATION PROCESS

The common types of fermentation processes that are applied for lactic acid production are the following: batch, continuous, fed batch and repeated batch. The proper selection of each type depends on the challenge each time. Particularly, the parameters that are evaluated are the produced lactic acid concentration, lactic acid productivity and lactic acid yield. Most of the fermentation processes are apllied through the batch mode. However, the implication of its type results in different outcomes. Specifically, the conduction of a batch fermentation system leads to higher lactic acid concentrations, the continuous fermentation systems lead to higher lactic acid productivities and finally higher lactic acid yields are obtained when fed batch and repeated batch systems are used. On the other hand, each fermentation mode has some general advantages and drawbacks. For example, a continuous system can operate for longer times compared to batch systems, while batch production of lactic acid through cheese whey fermentation exhibits long lag periods and unusually long fermentation times that require high fermenter capacity.

On the other hand, critical parameters should be taken into account such as, the optimum temperature of the fermentative strain applied, the addition of various nutrients for the enhancement of lactic acid production, the fermentation pH, that it is usually controlled and finally the avoid or decrease of by product formation.

Agroindustrial Wastes as Substrate for Lactic Acid Production

Lactic acid requirements are increasing annually. Thus, the exploitation of various raw materials through the fermentation process is very critical. The prerequisites regarding the selection of raw materials for lactic production are very strict. Specifically raw materials should be cheap, abundant and renewable, with low levels of contaminants and able to lead to high productivity, high yield, need minimum pre-treatment and form low levels of by-product.

On the other hand, the proper selection of the substrate depends on the geographical availability. For example, agriculture materials are preferable to expensive, pure sugars. In Northern Europe wheat is a major crop, whereas corn is used in the US.

Regarding the main agricultural wastes that have been applied for lactic acid production the following have been examined: molasses, sweet sorghum, starchy material (wheat, corn, cassava, and rice), cellulose, potato and cheese whey (Hofvendahl, Hahn–Hagerdal, 2000). On the other hand, hydrolysates from trimming vine shoots, barley bran husks, or corncobs have been also employed as carbon source, and complete bioconversion of hemicellulosic sugars was achieved (Moldes, Torrado, Converti, Domínguez, 2006).

Cheese Whey Fermentation for Lactic Acid Production

Although lactic acid can be produced by fermentation from a variety of sugar substrates, cheese whey is the predominant agricultural substrate that is exploitated for lactic acid production. A typical composition of cheese whey is presented at Table 2.

Cheese whey is the remaining liquid by-product following the precipitation and removal of milk casein during cheese-making. The gross composition of dried and liquid cheese whey is given in Table 3.

This by-product represents approximately 85-95% of the milk volume and retains 55% of milk nutrients. Among the most abundant of these nutrients are lactose, soluble proteins, lipids and mineral salts (Table 2). Several possibilities for cheese-whey exploitation have been examined over the last 40 years.

However, approximately 50% of world cheese-whey production is discarded as effluent and it is not treated. The rest is transformed into various food products. The two forms of cheese whey that are treated are (i) the liquid form, as powdered cheese whey, lactose and de-lactosed by-products, and (ii) the rest as cheese-whey-protein concentrates. Cheese whey disposal causes severe environmental problems because of the high volumes produced and its high organic matter content, exhibiting high values for BOD_5 (~30,000-50,000 ppm) and COD (~60,000-80,000 ppm), with lactose being largely responsible for the high BOD and COD levels (Marwaha and Kennedy, 1988; Gardner, 1989; Kemp and Quickenden, 1989; Mawson, 1994). A typical example about the negative impact of cheese whey on waste water of cheese manufacturing characteristics is presented in Table 4.

In order to reduce the BOD level and to acquire some useful compounds, this nutrient-rich whey can be utilized for the production of lactic acid by bacteria as a cheap carbohydrate source (Büyükkileci, Harsa, 2004). Several LAB and other microorganisms have been applied for lactic acid production through cheese whey in free or immobilized form in various matrixes. Regarding free LAB that have been applied for cheese whey fermentation, *Lactobacillus helveticus* was the most significant in terms of lactic acid production, lactic acid yield and lactic acid productivity.

Even though various LAB (*Lactobacillus helveticus*, *Lactobacillus delbrueckii* subsp. *bulgaricus*, *Lactobacillus acidophilus*, *Lactobacillus casei*, etc.) have been applied for the exploitation of cheese whey for lactic acid production, *Lactobacillus heleveticus* is the generally proper organism, as it produces approximately twice the amount of lactic acid from milk, compared to other common LAB (*Steptococcus thermophilus* and *Lactobacillus delbrueckii* subsp. *bulgaricus*) (Roy, Goulet, and LeDuy, 1986). Specifically, very high lactic acid production rate was recorded with *Lactobacillus helveticus* precultivated on hydrolysed whey supplemented with yeast autolysate (Amrane and Prigent, 1993). The control of temperature and pH value is crucious for the efficient lactic acid production process. *Lactobacillus helveticus* showed ameriolated lactose degradation and lactic acid production with increasing temperature (23–42 °C) exhibiting maximum lactic acid production at 42 °C (Tango and Ghaly, 1999). The effect of pH control range on the morphology of *Lactobacillus helveticus* was studied in batch cultures, and highest lactic acid productivity was obtained at pH 5.5 (Norton, Lacroix, and Vuillemard, 1993). Furthermore, during lactic acid production at high cell concentrations in whey ultrafiltrate by *Lactobacillus helveticus* the optimized fermentation temperature was 42 °C and the controlled pH value near 5.8 (Kulozik and Wilde, 1999).

Table 2. Typical composition of sweet and acid whey

Components	Sweet whey	Acid whey
	(g L^{-1})	(g L^{-1})
Total solids	63.0–70.0	63.0–70.0
Lactose	46.0–52.0	44.0–46.0
Protein	6.0–10.0	6.0–8.0
Calcium	0.4–0.6	1.2–1.6
Phosphate	1.0–3.0	2.0–4.5
Lactate	2.0	6.4
Chloride	1.1	1.1

Source: Marwaha and Kennedy, 1988.

Table 3. Gross composition of liquid and dried whey (Henning 1998)

Component (%)	Liquid whey	Dried whey
Total solids	6.35	96.5
Protein	0.8	13.1
Lactose	4.85	75.0
Fat	0.5	0.8
Minerals and vitamins	0.6	1.2
Lactic acid	0.05	0.2
Ash	0.5	7.3

Table 4. The characteristics of a typical waste water from cheese manufacturing with and without whey recovery

Parameter	Installation with whey recovery	Installation without whey recovery
mg L^{-1}		
BOD_5	2397	5312
COD	5312	20559
Fats	96	463
N total	90	159
P total	26	21

Source: http://eippcb.jrc.es/reference/fdm.htm). Integrated Pollution Prevention and Control Reference Document on Best Available Techniques in the Food, Drink and Milk Industries August 2006.

However, lactic acid bacteria have some nutritional requirements and especially their nitrogen requirements are multifarious (Chopin, 1993), (Pritchard and Coolbear, 1993). Cultivation media, which may have high protein content, are usually supplemented with yeast extract or protein lysates. In many fermentation studies, yeast extract is usually an essential nutrient for *Lactobacillus* for efficient lactic acid production (Amrane, 2005; Schepers et al, 2002). Few other nitrogen sources have proved to be as good in promoting the process (Hujanen and Linko, 1996). However the cost of the whole procedure is high therefore, supplementation using other nutrients, like molasses, corn steep molasses and amino acids to whey, has also been examined for increasing lactic acid yield (Liu et al, 2004). Indeed, the addition of molasses to whey has resulted in improvements in lactic acid production (Chiarini and Mara, 1990). In another study, high lactose consumption (94.1%) together with efficient lactic acid production (26.1 g L^{-1}) and yield (0.90%) were obtained in whey ultrafiltrate supplemented with 1% beet molasses (Chiarini, Mara, and Tabacchioni, 1992).

On the other hand the application of immobilization techniques for cheese whey valorization has been also examined. Cell immobilization is the physical confinement or localization of intact cells to a certain region of space with preservation of some desired catalytic activity (Karel *et al*, 1985). Several immobilization matrices have been applied as it indicated in Table 5. Immobilization techniques are gaining scientific interest during the last years due to many advantages over free cells. The main of them are altered growth rates, increased substrate uptake and product yield, lower yield of fermentation by-products, higher intracellular pH values and increased tolerance against toxic and inhibitory compounds.

Several chemical and physical techniques have been proposed for cell immobilization. Adsorption, gel entrapment, and covalent binding are the most popular methods of immobilization used in various bioprocesses (Table 6).

Among the supports and methods for cell immobilization, the most useful for food processing is the entrapment using natural polymers like alginate, agarose, carrageenan, chitosan and pectin. Such polysaccharides represent an emerging group due to various advantages such as: being non-toxic, biocompatible and cheap. Likewise, entrapment is considered as the most efficient for cheese whey fermentation. Through entrapment, the cells are allowed to penetrate into the porous matrix until their mobility is obstructed by the presence of other cells. In general, entrapment techniques are based on the inclusion of cells in a rigid environment to prevent the cells from diffusing into the surrounding medium and simultaneously allowing mass transfer of nutrients and metabolites.

Table 5. Different immobilization matrices used for lactic acid production from whey

Microorganism	Immobilization matrix
Lactobacillus casei	Agar-agar
Lactobacillus casei	Polyacrylamide
Lactobacillus delbrueckii subsp.*bulgaricus*	Hollow-fibre
Lactobacillus helveticus	Alginate
S. thermophilus	κ-Carrageenan/locust bean gum
S. thermophilus	κ-Carrageenan/locust bean gum
Lactobacillus delbrueckii subsp. *bulgaricus*	κ-Carrageenan/locust bean gum
Lactobacillus casei and *Lc. lactis*	Alginate
Lactobacillus casei and *Lc. lactis*	κ-Carrageenan
Lactobacillus casei and *Lc. lactis*	Agar-agar
Lactobacillus casei and *Lc. lactis*	Polyacrylamide
Lactobacillus casei	Agar-agar
Lactobacillus casei subsp. *casei*	Porous sintered glass beads
Lactobacillus delbrueckii subsp. *bulgaricus*	κ-Carrageenan
Lactobacillus helveticus	κ-Carrageenan/locust bean gum
Lactobacillus helveticus	Alginate
Lc. lactis subsp. *lactis*	κ-Carrageenan/locust bean gum
Lc. lactis subsp. *lactis* biovar *diacetylactis*	κ-Carrageenan/locust bean gum
Lactobacillus casei	Poraver beads
Lactobacillus brevis	Delignified cellulosic material
Lactobacillus casei	Wood chips, brick particles porous glass, and egg shell
Lactobacillus casei	Apple and quince pieces
Lactobacillus casei	Alginate–chitosan
Lactobacillus helveticus	κ-Carrageenan/locust bean gum
Lactobacillus casei	Pectate gel

Source: Panesar, Kennedy, Dina, Gandhi and Bunko, 2007.

Table 6. Comparison of different immobilization techniques

Characteristic	Cross-linking	Adsorption	Chelation	Covalent binding	Entrapment
Preparation	Inter.	Simple	Simple	Diff.	Inter.
Binding force	Strong	Weak	Inter.	Strong	Inter.
Retention of activity	Low	High	Inter.	Low	Inter.
Regeneration of carrier	Imposs.	Poss.	Poss.	Rare	Imposs.
Cost of immobilization	Interm.	Low	Low	High	Low
Stability	High	Low	Inter.	High	High
Applicability in food technology	No	Yes	No	No	Yes
Protection from microbial attack	No	Yes	Yes	No	Yes
Viability	No	Yes	Yes	No	Yes

Adapted from Kennedy and Cabral *1983*.

Entrapment of bacterial cells in natural polysaccharide gel beads leads to increase of cell density in continuous fermentations and may result in improved productivity. Immobilized cells have been successfully used in repeated batch fermentation. Among different supports (calcium alginate, κ-carrageenan, agar, and polyacrylamide gels) tested for co-immobilization of *Lactobacillus casei* and *Lactococcus lactis* cells, alginate proved to be the best matrix for the production of lactic acid from deproteinised whey (Roukas and Kotzekidou, 1991). The immobilization process protected the cells from adverse conditions and improved the yields of lactic acid.

In other scientific research *Lactococcus* strains immobilized separately in κ-carrageenan-locust bean gum gel beads was also carried out in continuous fermentations of whey permeate medium (Lamboley, Lacroix, Champagne, and Vuillemard, 1997). The fermentation system exhibited a high mechanical stability for the 53-day continuous fermentation and no strain became dominant, or was eliminated from the bioreactor.

Lactobacillus casei was immobilized by adsorption in a recycle batch reactor system and the overall productivity of the fermentation system was higher, in comparison with the batch process using free cells (Senthuran,

Senthuran, Hatti-Kaul, and Mattiasson, 1999). The enhancement in productivity in the recycle batch reactor was also accompanied by an increase in density of suspended cells.

Lactobacillus brevis cells immobilized by adsorption on delignified cellulosic material resulted in 70% yield (Elezi *et al*, 2003). The system showed high operational stability with 10 repeated batch fermentations without any loss in cell activity. In another study, *Lactobacillus casei* was immobilized by adsorption on apple pieces and it was used for 15 successive fermentation batches of whey and milk. These immobilized biocatalysts proved to be very effective and suitable for food grade lactic acid production, since the fermentation system was stable and the conversion of lactose to lactic acid high. Other researchers studied the *Bifidobacterium longum* immobilized in sodium alginate beads and on a spiral-sheet bioreactor for the production of lactic acid from cheese whey was evaluated (Shahbazi, Mims, Li, Shirley, Ibrahim, Morris, 2005). The results showed that *Bifidobacterium longum* immobilized in sodium alginate beads showed better performance in lactose utilization and lactic acid yield than *Lactobacillus helveticus*. Specifically when immobilized *Bifidobacterium longum* in the spiral-sheet bioreactor was tested, the lactose conversion ratio was determined about 69% and the lactic acid yield was about 0.51 g of lactic acid/g of lactose utilized were obtained.

Tango and Ghaly (2002) reported that using a 5 L immobilized packed bed of *Lactobacillus helveticus* bed bioreactor in a continuous system under micro-aeration (0.10 vvm) conditions and yeast extract (10 g L^{-1}) addition, maximum lactic acid concentration can be achieved. Specifically the results regarding yield and productivity were higher than those reported in the literature for immobilized cell systems.

Immobilized *Lactobacillus casei* NRRL B-441 in alginate–chitosan beads has led to higher lactic acid production and lower cell leakage compared with Ca alginate beads and these gel beads were used for five consecutive batch fermentations without any marked activity loss and deformation (Göksungur, Gunduz and Harsa, 2005). The proposed processes were further developed and improved regarding the immobilisation of *Lactobacillus casei* using Ca pectate gel and the respective fermentation cell system was found very stable during whey fermentation to lactic acid (Panesar, Kennedy, Knill and Kosseva, 2007). Specifically high lactose conversion (94.37%) to lactic acid and high concentration of lactic acid (32.95 g L^{-1}) was obtained.

References

Amrane, A. Analysis of the kinetics of growth and lactic acid production for *Lactobacillus helveticus* growing on supplemented whey permeate, *Journal of Chemical Technology and Biotechnology*, 2005, 80, 345–352.

Amrane, A., Prigent, Y. Influence of media composition on lactic acid production rate from whey by *Lactobacillus helveticus*, *Biotechnology Letters*, 1993, 15, 239–244.

Buenaventurada, P., Calabia, Z., Tokiwa, Y. Production of D-lactic acid from sugarcane molasses, sugarcane juice and sugar beet juice by Lactobacillus delbrueckii, *Biotechnology Letters*, 2007, 29, 1329–1332.

Büyükkileci, A. O., Harsa, S. 2004. Batch production of L(+) lactic acid from whey by *Lactobacillus casei* (NRRL B-441), *Journal of Chemical Technology and Biotechnology*, 2004, 79, 1036–1040.

Chiarini, L., Mara, L. Lactic acid production from whey or ultrafiltrate by *Lactobacillus helveticus*, *Food Biotechnology*, 1990, 4, 101–105.

Chiarini, L., Mara, L., Tabacchioni, S. Influence of growth supplements on lactic acid production in whey ultrafiltrate by *Lactobacillus helveticus*, *Applied Microbiology and Biotechnology*, 1992, 36, 461–464.

Chopin, A. Organization and regulation of genes for amino acid biosynthesis in lactic acid bacteria, *FEMS Microbiology Reviews*, 1993, 12, 21–38.

Datta, R., Tsai, S. P., Bonsignor, P., Moon, S., Frank, J. Technological and economical potential of polylactic acid and lactic acid derivatives, *FEMS Microbiology Reviews*, 1995, 16, 221–231.

Demirchi, A., Pometto, A. Enhanced production of D (2)-lactic acid by mutants of *Lactobacillus delbrueckii* ATCC 9649. *Journal of Indian Microbiology*, 1992, 11, 23– 8.

Elezi, O., Kourkoutas, Y., Koutinas, A. A., Kanellaki, M., Bezirtzoglou, E., Barnett, Y. A. Food additive lactic acid production by immobilized cells of *Lactobacillus brevis* on delignified cellulosic material, *Journal of Agricultural and Food Chemistry*, 2003, 51, 5285–5289.

Giurca, R., Levin, R. E. Optimization of the lactic acid fermentation of hydrolyzed cod (*Gadus morhua*) gurry with corn syrup as carbohydrate source, *Journal of Food Biochemistry*, 1993, 16, 277– 89.

Göksungur, Y., Gunduz, M., Harsa, S. Optimization of lactic acid Production from whey by *L casei* NRRL B-441 immobilized in chitosan stabilized Ca-alginate beads, *Journal of Chemical Technology and Biotechnology*, 2005, 80, 1282–1290.

Gregor, R. The scientific basis for probiotic stains of Lactobacillus. *Applied and Environmental Microbiology*, 1999, 60, 3763–3766.

Henning, D. R., Fermented By-Products, In: *Applied dairy microbiology*, edited by Elmer, H. Marth, James Steele (Marcel Dekker, Inc, New York, 1998) p. 251-258

Hofvendahl, K., Hahn–Hagerdal, B. L-lactic acid production from whole wheat flour hydrolysate using strains of *Lactobacilli* and *Lactococci*, *Enzyme and Microbiology Technology*, 1997, 20, 301–7.

http://eippcb.jrc.es/reference/fdm.html. Integrated Pollution Prevention and Control Reference Document on Best Available Techniques in the Food, Drink and Milk Industries August 2006.

Huang, L. P., Jin, B., Lant P., Zhou, J. Simultaneous saccharification and fermentation of potato starch wastewater to lactic acid by *Rhizopus oryzae* and *Rhizopus arrhizus, Journal of Biochemical Engineering*, 2005, 23, 265-276.

Hujanen, M., Linko, Y. Y. Effect of temperature and various nitrogen sources on *l*(+)-lactic acid production by *Lactobacillus casei*, *Applied Microbiology and Biotechnology*, 1996, 45, 307–313.

Hujanen, M., Linko, Y.-Y. Effect of temperature and various nitrogen sources on L(1)-lactic acid production by *Lactobacillus casei, Applied Microbiology and Biotechnology*, 1996,45, 307–13.

Karel, S. F., Libicki, S. B., Robertson, C. R. 1985. The immobilization of whole cells-engineering principles, *Chemical Engineering Science*, 1985, 40, 1321–1354.

Kennedy, J F., Cabral, J. M. S. (1983). Immobilized Living Cells And Their Applications. In: I. Chibata and Wingard (Eds.), Applied Biochemistry and Bioengineering, V.4, *Immobilized Microbial Cells* (Pp. 189–280). New York: Academic.

Kourkoutas, Y., Xolias, V., Kallis, M., Bezirtzoglou, E., Kanellaki, M. *Lactobacillus casei* cell immobilization on fruit pieces for probiotic additive, fermented milk and lactic acid production, *Process Biochemistry*, 2005, 40, 411–416.

Kulozik, U., Wilde, J. Rapid lactic acid production at high cell concentrations in whey ultrafiltrate by *Lactobacillus helveticus*, *Enzyme and Microbial Technology*, 1999, 24, 297–302.

Lamboley, L., Lacroix, C., Champagne, C. P., Vuillemard, J. C. Continuous mixed strain mesophilic lactic starter production in supplemented whey permeate medium using immobilized cell technology, *Biotechnology and Bioengineering*, 1997, 56, 502–516.

Litchfield, J. H. Microbiological production of lactic acid, *Advances in Applied Microbiology*, 1996, 42, 45-95.

Liu, A., Liu, Y., Liao, W., Wen Z., and Chen, S. Simultaneous production of nisin and lactic acid from cheese whey: Optimization of fermentation conditions through statistically based experimental designs, *Applied Biochemistry and Biotechnology*, 2004, 114, 627–638.

Marwaha, S. S., Kennedy, J. F. Whey pollution problem and potential utilization, *International Journal of Food Science*, 1988, 23, 323-336.

Melzoch, K., Votruba, J., Habova, V., Rychtera, M. Lactic acid production in a cell retention continuous culture using lignocellulosic hydrolysate as a substrate, *Journal of Biotechnology*, 1997, 56, 25–31.

Moldes, A. B., Torrado, A., Converti, A., Domínguez, J. M. Complete bioconversion of hemicellulosic sugars from agricultural residues into lactic acid by Lactobacillus pentosus, *Applied Biochemistry and Biotechnology*, 2006, 135, 219-227.

NNFCC Renewable Chemicals Factsheet: Lactic Acid.

Norton, S., D'Amore, T. Physiological effects of yeast cell immobilization: applications for brewing. *Enzyme Microbiology Technology*, 1994, 16, 365–375.

Norton, S., Lacroix C., Vuillemard, J. C. Kinetic study of continuous whey permeate fermentation by immobilized *Lactobacillus helveticus* for lactic acid production, *Enzyme and Microbial Technology*, 1994, 16, 457–466.

Panesar, P. S., Kennedy, J. F., Knill C. J., Kosseva, M. R. Applicability of pectate-entrapped lactobacillus casei cells for l(+) lactic acid production from whey, *Biotechnological Products and Process Engineering*, 2007, 74, 35–42.

Parmjit, S., Panesar, Kennedy, J. F., Gandhi, D. N., Bunko, K. Bioutilisation of whey for lactic acid production, *Food Chemistry*, 2007, 105, 1-14.

Pritchard, G. G., Coolbear, T. The physiology and biochemistry of the proteolytic system in lactic acid bacteria, *FEMS Microbiology Reviews*, 1993, 12, 179–206.

Roukas, T. and Kotzekidou, P. Production of lactic acid from deproteinized whey by coimmobilized *Lactobacillus casei* and *Lactococcus lactis* cells, *Enzyme and Microbial Technology*, 1991, 13, 33–38.

Roy, D., Goulet, J., LeDuy, A. Batch fermentation of whey ultrafiltrate by *Lactobacillus helveticus* for lactic acid production, *Applied Microbiology and Biotechnology*, 1986, 24, 206–213.

Samuel, W. A., Lee, Y. Y., Anthony, W. B. Lactic acid fermentation of crude sorghum extract, *Biotechnology and Bioengineering*, 1980, 22, 757–78.

Schepers, A. W., Thibault, J., Lacroix, C. *Lactobacillus helveticus* growth and lactic acid production during pH-controlled batch cultures in whey permeate/yeast extract medium. Part I. multiple factor kinetic analysis, *Enzyme and Microbial Technology*, 2002, 30, 176–186.

Senthuran, A., Senthuran, V., Hatti-Kaul R., and Mattiasson, B. Lactic acid production by immobilized *Lactobacillus casei* in recycle batch reactor: A step towards optimisation, *Journal of Biotechnology*, 1999, 73, 61–70.

Shahbazi, A., Mims, M. R., Li, Y., Shirley, V., Ibrahim, S. A., Morris, A. Lactic acid production from cheese whey by immobilized bacteria, *Applied Biochemistry and Biotechnology - Part A Enzyme Engineering and Biotechnology*, 2005, 122, 529-540.

Tango, M. S. A., Ghaly, A. E. A continuous lactic acid production system using an immobilized packed bed of *Lactobacillus helveticus*, *Applied Microbiology and Biotechnology*, 2002, 58, 712-720.

Tango, M. S. A., Ghaly, A. E. Amelioration of lactic acid production from cheese whey using micro-aeration, *Biomass and Bioenergy*, 1999, 17, 221-238.

Timbuntam, W., Sriroth, K., Tokiwa, Y. Lactic acid production from sugar-cane juice by a newly isolated Lactobacillus sp, *Biotechnology Letters*, 2006, 28, 811–814.

Tiwari, K. P., Pandey, A., Mishra, N. Lactic acid production from molasses by mixed population of Lactobacilli. Zentralbl Bakteriol Parasitenkd Infektionskr Hyg Zweite Naturwiss Abt Mikrobiol Landwirtsch Technol Umweltschutzes, 1979, 134, 544–6.

Vijayakumar, J., Aravindan, R., Viruthagiri, T. 2008. Recent trends in the production, purification and application of lactic acid. *Journal of Chemical Technology and Biotechnology*, 2008, 22, 245–264.

Wang, M., Xu, Z. J., Qiu, T. L., Sun, X. H., Han, M. L., Wang X. M. Kinetics of Lactic Acid Fermentation on Food Waste by *Lactobacillus bulgaricus, Advanced Materials Research*, 2010, 113-116, 1235.

Wu, X., Jiang, S., Liu, M., Pan, L., Zheng, Z., Luo, S. Production of L-lactic acid by *Rhizopus oryzae* using semicontinuous fermentation in bioreactor, *Journal of Industrial Microbiology and Biotechnology*, 2011, 38, 565-571.

In: Lactic Acid
Editors: A. Migallón and R. Ruseckaite
ISBN: 978-1-62081-643-1

Chapter 7

TUNING CELL BEHAVIOR IN POLY (LACTIC ACID) BASED NANOCOMPOSITE SCAFFOLDS

I. Armentano,[1] E. Fortunati,[1] S. Mattioli,[1] N. Rescignano[1,2] and J. M. Kenny[1,2]
[1]Materials Engineering Center, UdR INSTM, University of Perugia, Terni, Italy
[2]Institute of Polymer Science and Technology, ICTP-CSIC, Madrid, Spain

ABSTRACT

This chapter reports the relevance and the promising perspectives of poly(lactic acid) (PLA) based nanostructured biomaterials in tissue engineering. The development of PLA matrix nanocomposites with different morphologies and the effects of their surface nanotopography are here extensively investigated in terms of polymer physicochemical properties and stem cell response. Aliphatic polyester polymers, such as PLA, have already attracted wide attention for the preparation of biomaterials taking advantage of their biodegradability and biocompatibility in the human body. Polymeric nanocomposite scaffolds represent a valuable strategy that offers the opportunity to modulate the surface, mechanical and chemical properties of polymer by adding a specific type and content of nanostructure to the biomaterial. The possibility to control cell function by modulating the polymer properties represents a key point of material science in tissue engineering applications. Combining surface research efforts from the physical and chemical approaches may lead to a wide range of cellular responses and activities. Nanostructured biodegradable polymeric materials are ready for take off and certainly promise an exciting future at the interface of chemistry, biology and material science.

Keywords: Nanocomposite; Scaffold; Topography; Mechanical Properties; Electrical Properties

INTRODUCTION

The nanocomposite approach has emerged in the last two decades as an efficient strategy to upgrade the structural and functional properties of synthetic polymers by the combination of polymers with organic/inorganic nanostructures [1,2]. The design and preparation of multi-component polymer systems hence has permitted to develop innovative multifunctional biomaterials, with specific properties able to tune cell behaviour [3]. Aliphatic polyesters play a very important role in the field of biodegradable materials and now their nanocomposites are attracting growing interest from researchers. Nanotechnology and the hereby produced nanostructures, such as layered silicates, layered titanates, carbon nanotubes, gold, silver, maghemite and magnetite nanoparticles, nanocellulose, etc. have promised to make use of specific properties of such supramolecular assemblies and materials so that hitherto breakthrough effects can be exploited for new applications [4].

Due to its biocompatibility and high mechanical properties, poly(lactic acid) (PLA) has been approved for different in vivo bioapplications. As it degrades to natural metabolites, it has been the predominant choice for materials in sutures and biodegradable implantable meshes for guided tissue regeneration [5,6]. The basic building block for PLA is lactic acid, which is a simple chiral molecule that exists as two enantiomers, L- and D-lactic acid, differing in their effect on polarized light. Figure 1 represents a scheme of lactic acid interaction with living systems: starting from the biopolymer synthesis, nanocomposite approach, scaffold development, stem cell interaction and concluding with tissue regeneration.

The development of nano-biocomposites based on poly(lactic acid) is a convenient approach to nanostructured materials, which includes the use of nanoparticles, nanofibres, and other nanoscaled features, mainly in polymer or bioceramic matrices [7]. The key premise in tissue engineering strategies, however, is the use of biodegradable polymers to make three-dimensional (3D) porous scaffold structures which allow entrance, adhesion and proliferation of regenerating cells. PLA macroporous or fibre-based scaffolds and PLA highly porous sponges have been fabricated to obtain synthetic extracellular matrices (ECM) employed to engineer organic tissues. Various cell types have been cultured on these fibrous scaffolds and successfully transplanted into animal models [8]. Nanocomposite scaffolds can be produced by casting, salt leaching, dry freezing, CO_2 gas foaming, electrospinning, etc. combining biodegradable polymers with functional nanostructures able to transfer their specific properties to the biodegradable polymer [9–11]. Based on these considerations, in this chapter we have focused our attention on the bulk and surface nanostructure of PLA matrix nanocomposites, on their functional properties and on their role in cell attachment and proliferation, summarizing also some our recent research activities in this field [1,7,8,11].

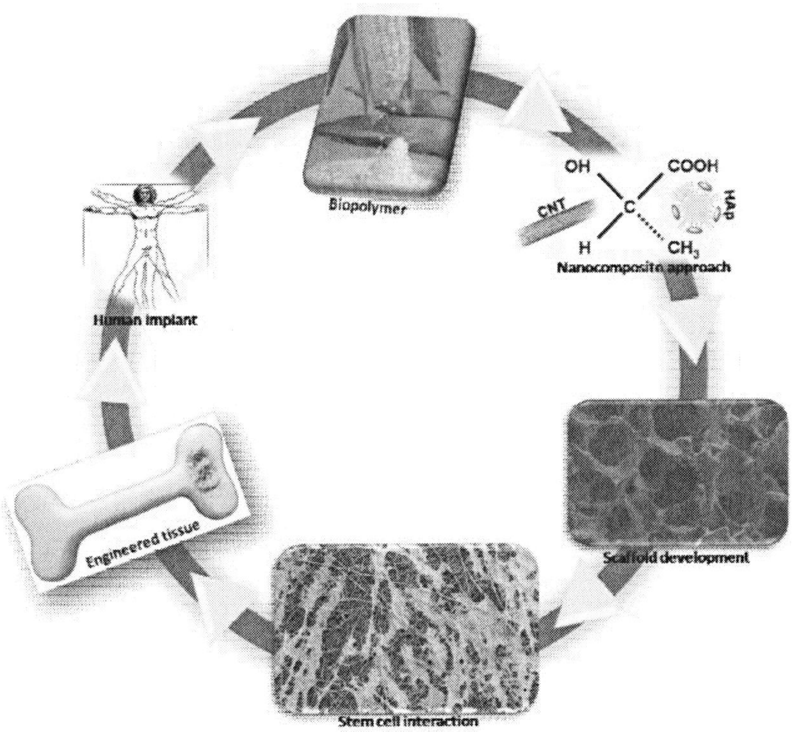

Figure 1. Lactic acid interaction with living systems: a tissue engineering approach.

NANOSTRUCTURES

Different kinds of organic and inorganic nanostructures (carbon nanostructures, metal, ceramic nanoparticles and polymeric nanoparticles, cellulose micro- and nano-crystals, etc.) have been used as fillers in PLA based nanocomposite scaffolds in order to modulate properties that are known to affect and tune cell behaviour [1]. Nanostructures can be divided based on their chemistry or on their morphology. They are classified as 0D, 1D, 2D according to the number of their nanometric dimensions. Conventionally, spherical nanoparticles are defined 0D, nanotubes and nanofibres as 1D and platelets (graphene, nano clays, etc.) as 2D [12]. Following this approach, we will introduce a brief description of the main nanostructures that are currently used as fillers in PLA for tissue engineering applications.

Organic Nanostructures

Carbon nanostructures (CNS) as nanofibres (CNFs), nanotubes (CNTs), graphene (GS), graphene oxide (GO) and nanodiamond (ND) are considered to be ideal reinforcing agents for high-strength polymer composites, because of their tremendous mechanical strength, high electrical conductivity and high aspect ratio [13]. They are used in order to modulate specific mechanical, thermal and electromagnetic polymeric properties, taking also advantage of their versatile potential applications [14]. Carbon nanofibres are cylindrical or conical structures with diameters varying from a few to hundreds of nanometres and lengths ranging from less than a micron to millimetres. The internal structure of carbon nanofibres is comprised of different arrangements of modified ordered graphene sheets [13]. Excellent electrical conductivity and mechanic-chemical properties characterize CNFs, as suitable candidates for polymer nanohybrid materials, since the nanosized diameter (i.e., 70-200 nm) provides high specific areas with enormous potential to modify the polymer matrix properties and, in particular, to improve the electrical conductivity and the mechanical properties [15]. Carbon nanotubes can be classified in single walled carbon nanotubes (SWCNTs) and multi walled carbon nanotubes (MWCNTs). A SWCNT is represented by a single graphene sheet rolled into a tube with elastic modulus up to 1 TPa and a tensile strength close to 60 GPa which signifies 5- and 10-times greater elasticity and strength than steel at just 1/6 of its weight [16], MWCNTs consist in more graphene sheets rolled concentrically into a cylinder. The exceptional physical properties, as well as their high aspect ratio, make carbon nanotubes ideal candidates for the next generation of polymer composites [17,18]. Their electrical properties (metallic or semiconducting behaviour) depend on the chiral angle, i.e., the angle at which the atoms are twisted on the main axis of the carbon nanotube itself [19]. SWCNTs display several unique chemical, electrical, and mechanical properties that make them interesting nanomaterials for the development of innovative biomedical devices. However, SWCNTs efficacy is still debated and controversial, especially in terms of cytotoxicity. Several studies, in fact, have been conducted with different results. For example, SWCNTs resulted to be cytotoxic for alveolar macrophages [20] and human fibroblasts [21,22]. ND is composed of particles with about 5 nm average diameter consisting of the inert diamond core terminated with surface functional groups such as COOH, OH, NH_2, etc. [23].

Polymeric nanoparticles based on biodegradable polymers have recently attracted a strong interest since they may provide a convenient and non-invasive method for the administration of genes or drugs and also facilitate sustained site directed delivery to specific disease cell types or tissues bypassing the obstructive pathophysiological barriers [24, 25]. Particle size is one of the key features in determining performance in biomedicine applications because it influences circulating half-life, cellular uptake and biodistribution, and the drug release kinetic [26–28]. The uniformity of the particle population is also a significant factor in performance. Particle preparations procedure lacking uniformity will exhibit variance in all of these parameters, making it difficult to draw conclusions about which subset of the particle population is responsible for biological effects [29]. There are several techniques potentially useful for the preparation of biodegradable polymeric nanoparticles, such as: emulsion-based methods, evaporation or extraction of solvent, nanoprecipitation, salting out, spray-drying method, etc. [28,30,31]. The selected method determines the characteristics of spheres, including the size, the entrapment of drugs, the administration mode and the drug release [28–30]. Polymer degradation can also be affected by the particle size and can have an important role in the drug release kinetics.

Inorganic Nanostructures

Calcium phosphate-based bioceramics, such as hydroxyapatite (HA), $Ca_{10}(PO_4)_6(OH)_2$, β-tricalcium phosphate (β-TCP), $Ca_3(PO_4)_2$, and biphasic calcium phosphate (BCP), a mixture of HA and β-TCP that is composed of the same ions as bone, are the inorganic materials which have received most attention for bone repair applications. HA is the major inorganic constituent (69 %wt.) of the bone and it is the prospective one of inorganic biomaterials for bone regeneration. Synthetic HA is biocompatible, osteoconductive and osteoinductive and it is thus widely employed for hard tissue repair in orthopaedic and dental applications developing a mechanically tight bond with bone [32,33]. It has been shown, in fact, that sintered hydroxyapatite bind serum proteins and promotes the adsorption of adhesion ligands and integrins responsible for cell adhesion to the surface. HA ceramics are high biocompatible with bone tissue and have a high osteogenic potential (osteoconduction and osteointegration), but the brittleness and fatigue failure in the body limit their clinical applications only for unload bearing repair and substitute. Another outstanding problem of HA ceramics is their bad biodegradability. Sintered HA materials showed no detectable resorption over a period of 9 months of implantation. Especially, nano-HA is of current interest because it possesses some special properties, including large specific surface area, improved biodegradability and biological activity. When compared to β-TCP, HA resorbs slowly and undergoes little conversion to a bone-like material after implantation. However, for the same porosity, β-TCP scaffolds often have lower strength than HA scaffolds, so their use in the repair of load-bearing bones may be challenging. The use of BCP with different HA to β-TCP ratios allows manipulation of the degradation rate [34], as well as other properties [35,36]. Furthermore, it is well known that Ca-deficient-hydroxyapatite (d-HA) presents features similar to biological apatite with respect to stoichiometric hydroxyapatite [37–39]. Moreover, d-HA has a greater bioresorption degree and elicits the formation of an apatite layer in physiological solutions [40].

Nanoparticles of noble metals are also considered as fillers for poly(lactic acid) since they exhibit significantly distinct physical, chemical and biological properties, strongly influenced by shape and size [41,42]. The goal is to obtain small particle sizes, narrow size distributions and well stabilized metal particles. Because of surface effects and the dramatic changes in properties occurring when the critical length, which governs some physical phenomenon (magnetic, mechanical, etc.) becomes comparable with size, metal clusters have unique properties (e.g. plasmon absorption, near-IR photoluminescence, super-paramagnetism, etc.). The embedding of nanoscopic metal structures into polymeric matrices represents the simplest way to protect clusters and take advantage of their physical characteristics [43].

PROCESSING OF PLA NANOCOMPOSITE SCAFFOLDS

The ideal scaffold should (i) be biocompatible (not toxic) and should promote cell adhesion and proliferation; (ii) exhibit, after in vitro tissue culture, mechanical properties that are comparable to those of the tissue to be replaced; (iii) have a porous three-dimensional (3D) architecture to allow cell proliferation, vascularization and diffusion of nutrients between the cells seeded within the matrix and the surroundings; (iv) degrade at a rate that matches the production of new tissue, into non-toxic products that can be easily resorbed or excreted by the body; and (v) be capable of being processed economically into anatomically relevant shapes and dimensions, and be sterilized for clinical use [7,44,45]. Scaffolds are crucial to tissue engineering strategies for a number of reasons; as a three-dimensional structure they provide volume fill, mechanical integrity and a surface that can provide chemical and architectural guidance for regenerating tissues. Scaffolds provide a matrix for guided cell proliferation and should be able to control the shape of the new formed tissue. A high porosity and a high pore interconnectivity of a scaffold are desired to minimize the amount of implanted material and to increase the specific surface area for cell attachment and tissue ingrowth [45,46]. Research in this area has been dominated by developing new scaffold materials and fabrication techniques. Scaffolds for tissue engineering are commonly constructed from biodegradable synthetic or natural polymeric materials [46]. Scaffold morphology, in terms of interconnectivity, pore size, shape and surface physical and chemical properties is a key point in stem cells interaction. Several methods have been developed to fabricate porous biodegradable nanocomposite scaffolds including electrospinning [11,47,48], solvent casting/particulate leaching [49], particle sintering [50], three-dimensional printing [51], gas foaming [52,53], and phase separation [54].

Electrospinning

Electrospinning is an easy and versatile technique for nano- and micro-fibre fabrication, which can produce continuous fibres ranging in diameter from several micrometres to 100 nm or less by adjusting the fabrication parameters [11,47,48]. The process is based on the application of a high-voltage electric field to a polymeric solution jet delivered through a nozzle. As soon as the applied electric field overcomes the surface tension of the droplet, a continuous polymer filament is drawn by high electrostatic forces and deposited onto a grounded metallic target. In order to meet different requirements for specific applications in tissue engineering, nanofibrous scaffolds with defined diameters, pore sizes and porosities, patterns and alignments have been successfully prepared by electrospinning [48,55,56]. Electrospinning technology can be used to fabricate nonwoven nanofibrous scaffolds from biological and/or synthetic polymers and has tremendous potential for tissue engineering applications [11,57]. The diameter of the individual fibres can be specifically controlled down to the nanometer range, and the fibres can be patterned through a variety of methods. Moreover, the electrospun nanofibres have a large surface area to volume ratio, which allows for the direct attachment of ECM ligands and growth factors onto fibre surfaces to locally modulate cell and tissue function and to enhance regeneration [58,59].

In order to improve the potential applications of electrospun scaffolds, the addition of specific fillers has been positively considered. The electrospinning of PLLA/ceramic hybrids can be regarded as an effective technique, able to form ECM resembling mats containing specific fillers to improve their mechanical characteristics and to furnish signalling cues to the seeded cells. Moreover, many authors reported that electrospun hybrid scaffold based on bioresorbable polymers and hydroxyapatite allow osteoblast proliferation and differentiation, and are thus considered very promising as bone scaffolding materials [11,57,58,60].

Gas Foaming

Gas foaming is a promising technique, selected as a solvent-free process, which allows functional fillers to be added to a porous polymeric matrix [52,53]. A supercritical fluid (SCF) is created once a substance is exposed to an environment where its critical temperature and pressure (critical point) are exceeded. Under these conditions, the liquid and gaseous components of the material become identical, and further compression of this fluid phase will not result in liquefaction. SCFs combine the properties of the two phases from which they are formed; they have densities and solvating properties similar to those of liquids, but in combination with a diffusivity and viscosity comparable with that of a gas. The solvent properties of these materials can in fact be controlled very precisely by small manipulations in the pressure (and therefore density) at with the SCF is processed [61]. CO_2 is the most common candidate for use as a SCF due to its low toxicity, flammability and cost, ready availability, stability, and environmental acceptability. In addition, the critical point conditions of 31 °C and 73.8 bars are easily attainable. This foaming process can be used to develop polymer, blend and nanocomposite foams with specific properties. Pore size and overall porosity were shown to decrease when fillers or fibres were added to a polymer matrix [62,63]. PLA foams reinforced with continuous glass fibres were processed by Bhler et al. [63] with a fibre fraction, V_f ranged from 0 to 15 %vol. and porosity ranged from 50 % to 92 %. With up to 1.5 GPa these cellular composites showed a higher longitudinal compression modulus than previously suggested foams made of bioresorbable materials. In addition, the elastic collapse strength increased with V_f and density. It is thus possible to tailor scaffolds with precise structural and mechanical properties. Open cell composite foams made of bioresorbable PLA and ceramic fillers, hydroxyapatite (HA) or β-tricalcium phosphate (β-TCP), were processed by supercritical CO_2 foaming [62]. Furthermore compressive tests demonstrated anisotropy in mechanical behaviour, with an axial modulus up to 1.5 times greater than the transversal modulus. Composite scaffolds also showed viscoelastic behaviour with increased modulus for higher strain rates. Such scaffolds prepared by gas foaming of polymer composite materials therefore possess suitable morphology and properties for bone tissue engineering applications.

Solvent Casting

The solvent casting/particulate leaching (SCPL) method is the most commonly process used for scaffold development. This approach allows for the preparation of porous structures with regular porosity, but with a limited thickness. The method is characterized by its simple operation and adequate control of the pore size and the porosity by selecting the soluble particle size and the amount of the added salt particles. First, the polymer is dissolved into a suitable organic solvent (e.g. poly(lactic acid) can be dissolved into dichloromethane), then the solution is cast into a mould filled with porogen particles. Such porogen can be an inorganic salt like sodium chloride, crystals of saccharose, gelatine or paraffin spheres, etc. The size of the porogen particles will affect the size of the scaffold pores, while the porosity of the final structure is directly correlated to the polymer to porogen ratio [49]. After the polymer solution has been cast, the solvent is allowed to fully evaporate, then the composite structure in the mould is immersed in a bath of a liquid suitable for dissolving the porogen: water in the case of sodium chloride, saccharose and gelatine or an aliphatic solvent like hexane for use with paraffin [8]. Once the porogen has been fully dissolved, a porous structure is obtained. Other than the small thickness range that can be obtained, another drawback of SCPL lies in the use of organic solvents which must be fully removed to avoid any possible damage to the cells seeded on the scaffold. Moreover, the distribution of the salt particles is often not uniform within the polymer solution. This is because the density of the liquid polymer solution and that of the solid salt are different, and the degree of direct contact between the salt particles is not well controlled. As a result, the interconnectivity of pores in a final scaffold cannot be well controlled. Furthermore, the polymer solution and the salt particles are mixed in such a way that the salt particles tend to be wrapped completely by the polymer. These wrapped salt particles cannot be easily leached out with water. Thus, most porous materials prepared by the SCPL method are limited to a thickness lower than 4 mm.

Thermal Induced Phase Separation (TIPS)

Similar to the previous technique, this phase separation procedure requires the use of a solvent with a low melting point that can easily sublime. For example, dioxane could be used to dissolve poly(lactic acid), then phase separation is induced through the addition of a small quantity of water: a polymer-rich and a polymer-poor phase are formed. Following cooling below the solvent melting point and some days of vacuum-drying to sublime the solvent, a porous scaffold is obtained. Liquid-liquid phase separation presents the same drawbacks of emulsification/freeze-drying [54]. TIPS process is currently used to fabricate porous scaffold for tissue engineering. This technique is based on the principle that a single homogeneous polymer solution made at an elevated temperature is converted via the removal of thermal energy to a polymer-rich and a polymer-lean phase. The polymer solution is quenched below the freezing point of the solvent to produce a porous structure. The pore sizes and their interconnectivity in the fabricated foams can be adjusted by modifying the thermodynamic and kinetic parameters, such as polymer concentration, quenching time, temperature, etc. [64].

CELL BEHAVIOUR MODULATION

For tissue engineering applications, PLA nanocomposites have to be specifically designed and developed in order to provide informative microenvironment that will allow cells to interpret the biomaterial instructions and modify their fate accordingly [65]. In particular, based on its composition and structure, the biomaterial must transmit specific stimuli to the cells that will decode these into biochemical signals. Hence, topography, chemistry and physical properties of biomaterials are critical parameters for directing cell fate [66]. In this perspective, we will focus on the properties of PLA that could be modulated by nanostructure introduction and that could influence cell behaviour. In particular, morphology, mechanical and electrical properties will be addressed, underlining the main results of our recent research activities.

Effects of PLA Nanocomposite Mechanical Properties

Poly(lactic acid) has been investigated for bone scaffold because of the good biocompatibility and biodegradability of this polymer. However, it has insufficient mechanical strength [67]. Incorporation of filler particles into PLA is a promising approach to yield a composite material with improved mechanical properties. When the mechanical properties of a single-phase system are insufficient for a given application, nanocomposite materials offer an important strategy to impart increased strength and toughness. By introducing nanoscale filler components, a more effective enhancement of the mechanical properties can be obtained compared with traditional macro- and micro-scale fillers, due to the greater surface area and lower percolation thresholds stemming from the higher aspect ratios of the nanoscale fillers [68]. Currently, most of the research on PLA is focused on composites made of a biopolymer and ceramic nanofillers with the aim to increase the mechanical properties and improve the interaction with surrounding tissues. For example, Deng et al. [69] produced nanocomposites of poly(lactic acid) with Ca-deficient hydroxyapatite nanocrystals (d-HA) through a solution casting technique. Nanocomposites with 10.5 %vol. d-HA showed a 49 % increase in the modulus as compared to the neat poly(lactic acid). Wilberforce et al. [70] have shown that by incorporating either HA microparticles or nanoparticles into poly (L-lactic acid) (PLLA) matrix the thermal and dynamic mechanical properties of PLLA can be altered. It has been shown that these properties are highly dependent on the varying surface areas of HA particles and on the number of particles per unit volume in the composites. For example, the lower glass transition temperature of the nanocomposites than those of the microcomposites was thought to be due to the higher surface area of HA nanoparticles in PLLA, resulting in a greater interfacial area in the nanocomposites that allow a better polymer chain mobility at lower temperatures. The cold crystallization temperature of the nanocomposites were lower than those of the microcomposites due to the nanoparticles acting as more effective nucleating agents in PLLA for crystallite formation at lower temperatures. In general, in the nano-regime surface areas are much more important, but microparticle surface areas are so small that slight variations in particle number and size may not be significant. The storage modulus (E_0) values at 37 °C of the nanocomposites were higher than those of the microcomposites which approached the lower range of E_0 of cortical bone [70]. HA materials are very advantageous to be used in hard-tissue replacement composites. However, due to the brittleness of the HA and to the lack of interaction with polymer, the ceramic nanoparticles may present deleterious effects on the mechanical properties, when added at high loadings.

The interface adhesion of HA particles and polymer matrix plays a very important role and it represents the major factors affecting the properties of the PLA/HA composites. In order to increase the PLA/HA interfacial strength, various methods have been tried in the past [71–73]. In order to improve the bonding between hydroxyapatite particles and PLLA, and hence to increase the mechanical properties of the PLLA/HA composites, the HA nanoparticles were surface-grafted (g-HA) with the polymer and further blended with PLLA [74]. A 3D porous scaffold of poly(lactic-co-glycolic acid) (PLGA) and PLLA grafted nano-hydroxyapatite (g-HA) has been prepared with the solvent casting/particulate leaching method. The grafted HA nanoparticles appear more uniformly distributed on the surface of PLGA matrix than the un-grafted HA. The intramuscular implant study showed that there were two outcomes for the nanocomposite of PLGA/g-HA, mineralized or biodegraded completely, which may be up to the content of HA in polymer matrix and the balance of biodegradation, mineralization and tissue regeneration. As bone replacement, the scaffold of PLGA/g-HA exhibited rapid and strong osteoconductivity for repairing critical radius defects. The incorporation of bone morphogenetic factor (BMP-2) could enhance the osteogenic process of the composite implant. The authors concluded that the composite scaffold of PLGA/g-HA exhibited the similar strong ability of mineralization, osteogenesis and biodegradation in intramuscular implants and replacement for bone defects although the levels of Ca and P exposure and the actual content of HA in PLGA/g-HA are lower than that in PLGA/HA [12].

Further, Oksman et al. [75] used 40%wt of flax fibres and found 114.7% increase in the modulus of PLLA. Nanodiamond (ND), with its nanoscale particle size, showed superior physico-chemical properties, and excellent biocompatibility, and it is currently considered as a promising material for biomedical applications. In order to fully benefit from the advantages of ND as nanoparticulate filler for polymers, the quality of the filler dispersion in the matrix must be addressed since it determines the surface area of the nanoparticles available for interaction with the matrix and ultimately translates into mechanical properties of the resulting composite. Zhang et al. showed that a uniform dispersion of nanodiamond in the polymer led to significant increase in hardness and Young's modulus of the composites, resulting in more than 200% increase in Young's modulus and 800% increase in hardness, bringing the nanocomposite properties close to that of the human cortical bone [76]. Testing nanocomposites based on PLLA and modified ND, as a matrix supporting murine osteoblast cell growth for up to 1 week, showed that the addition of nanodiamond had no negative effects on cell proliferation. Nanodiamond serves as a multifunctional additive providing improved mechanical properties, bright fluorescence, and options for drug loading and delivery via surface modification. Intense fluorescence of these nanocomposite scaffolds can be used to monitor bone re-growth replacing the implant in vivo [76].

In a recent study we analysed the properties of PLLA nanocomposites based on pristine and carboxylated single-walled nanotubes [7], by studying respectively, the impact of SWCNTs and their functional groups (COOH) on the polymer matrix. The roles of SWCNT incorporation and COOH groups were also investigated in terms of biocompatibility by using human bone cells in adhesion and proliferation studies. Pristine (SWCNTs) and carboxylated (SWCNTs-COOH) carbon nanotubes were considered in order to control the interaction between PLLA and nanotubes. SWCNTs and SWCNTs-COOH showed a good dispersion in the polymer matrix and improved the PLLA crystallinity. Thermal, morphological and dynamic-mechanical analyses revealed that carboxylic groups on the tube sidewalls increased compatibility between PLLA and nanostructures. Mechanical properties demonstrated an enhancement related to the introduction and functionalization of carbon nanotubes. Biological investigations showed that osteoblasts cultured on PLLA/SWCNTs-COOH nanocomposites have higher cell adhesion and proliferation than osteoblasts cultured on PLLA and PLLA/SWCNTs nanocomposites. These studies suggested that the combination of biodegradable polymers and SWCNTs opens a new perspective in the self-assembly of nanomaterials and nanodevices for biomedical applications with tuneable properties. We concluded that substrate rigidity and nanotube dispersion could play an important role in the controlling processes involved in cell adhesion and proliferation and that PLLA/SWCNTs-COOH showed good biocompatibility and appeared to be a promising biomaterial candidate in promoting bone regeneration.

3D nanocomposite scaffolds, produced by electrospinning process, were also studied in our laboratories by using d-HA. All of the electrospun hybrids consisted of micro and sub-micro randomly oriented fibres. The PLLA loaded with 1 %wt. d-HA showed the most homogeneous microstructure. The d-HA contents affected the viscoelastic properties of the electrospun mats and induced a decrease of the storage modulus. The biological evaluation was performed on PLLA loaded with 1 %wt. of micro and sub-micro d-HA. Both samples cultured with murine bone marrow stem cells (mBM-MSCs) showed comparable viability and cell-fibre interaction. Thus, these electrospun mats are potential substrates for culturing primary mBM-MCSs [11]. In a recent study, we demonstrated that the combination of suitable architecture and composition of PLLA/d-HA electrospun composite mats produce a microenvironment that drives human stem cells toward osteogenic lineage in the absence of exogenous soluble osteogenic inducers. By contrast, in these experimental conditions, neat PLLA mats preserve the undifferentiated status of multipotent and pluripotent stem cells [59]. We suggest that this biological phenomenon might be the result of the modulation of nanocomposite properties. Addition of d-HA to PLLA affects the topography, the chemistry and the mechanical properties of the polymer matrix, as showed by the reduction on the viscoelastic modulus compare to neat PLLA. We reported that differences in storage and loss modulus could exert a direct effect on stem cell osteogenic differentiation that was independent of the stem cell type. A normal tissue cell not only applies forces but also, as reviewed here, responds through cytoskeleton organization (and other cellular processes) to the resistance that the cell senses, regardless of whether the resistance derives from normal tissue matrix, synthetic substrate, or even an adjacent cell [77]. The cell, in turn, is shown to respond to the resistance of the substrate, by adjusting its adhesions, cytoskeleton, and overall state. Although considerable attention has been directed at the responsiveness of individual cells to external forces (outside-in) that range from fluid flow to direct stretching and local twisting, we are now beginning to understand that cellular responses to cell exerted forces involve a feedback loop in that couples to the elasticity of the extracellular microenvironment. Molecular mechanisms involved in cellular responses to matrix stiffness are still open to investigation, mechanobiology is a broad field. Emphasized here is the recent recognition that most tissue cells not only adhere to but also pull on their microenvironment and thereby respond to its stiffness in ways that relate to tissue elasticity.

Effects of PLA Nanocomposite Electrical Properties

Electrical properties of the PLA based nano-biocomposite substrates are important issues in cell/scaffold and cell/cell interactions. Although the induced cell adhesion, migration, and orientation in response to electrical stimulus have been well documented using 2D cultured cells, electrically induced cellular behaviours in 3D scaffold remain still unknown. Moreover, an optimal application of electrical stimulus could offer a novel engineering technique to regulate cell type dependent cellular shape and orientation that are known to be involved in cell differentiation and growth [78]. The electrical properties of poly(lactic acid) can be modulated by introduction of conductive nanofillers (CNTs, CNFs, GO, and metal nanoparticles, etc.) [79]. In order to obtain a conductive nanocomposite, carbon nanotubes must form a three dimensional network in the polymer above a critical concentration, known as percolation threshold (PT), where a transition from non-conducting to conducting state occurs [80]. The PT depends on the nanostructure dispersion, as well as on the nanofiller aspect ratio leading to lower PT with the increase of the aspect ratio. Due to the carbon nanotube needle-like shape and high aspect ratio, their percolation threshold in polymer matrix composites is achieved at low concentrations ranging from 0.0025 to 4 %wt. [81], allowing the modulation of the polymer electrical properties without strong changes in other important aspects such as processability. Supronowicz et al. [82] used a current-conducting PLA/CNT composite substrate to deliver electrical stimulation to osteoblasts in vitro and provided cellular/molecular evidences. They found that alternating current electrical stimulation promotes various important osteoblast functions, such as cell proliferation, expression of genes for collagenous and non-collagenous proteins, and calcium deposition in the extracellular matrix. These functions are responsible for the chemical composition of the organic and inorganic phases of the bone matrix and are required for new in vivo bone formation. In fact, such functions have major consequences for bone repair, healing, and regeneration and may be responsible for the accelerated bone healing observed in animal models under electrical stimulation. The in vitro research exemplifies novel biomaterial approaches and alternative strategies for enhancing osteoblast functions that, although extremely promising, remain as yet unexplored for bone regeneration purposes in either clinical or tissue engineering applications.

The possibility to incorporate in a polymer matrix two different nanostructures with unique properties is a versatile strategy to transfer and to integrate in a single material two specific properties, enabling to realize multifunctional composites, for use in advanced applications like catalysis, energy storage, nano-biotechnology, etc. [67,83]. In particular, an original approach is given by the combination of single wall carbon nanotubes and hydroxyapatite in a PLLA polymer matrix [44]. The new designed materials might possess the unique properties offered by the synergistic interaction of carbon nanotubes and HA. From the results of this study it can be concluded that both a plain PLLA scaffold and a porous composite scaffold, containing CNTs and HA, showed the same in vivo release properties for BMP-2. The bone morphogenetic factor was released in a burst-like manner, followed by a sustained release without altering the bioactivity of the BMP-2. Incorporation of CNTs and HA did not prolong BMP-2 retention in the composite material [44].

Significant efforts have been devoted to fabricate various biomaterials to satisfy specific clinical requirements. Recently researchers have employed the electrospinning technique in the incorporation of multiwalled carbon nanotubes/hydroxyapatite (MWCNTs/HA) nanoparticles into PLLA and the fabrication of a composite membrane to satisfy the specific requirements of guided tissue regeneration. This work represents the first trial on the fabrication of a biomedical membrane which possesses dual biological functions [67]. This new type of membrane shows excellent dual biological functions and satisfied the requirement of the guided tissue regeneration (GTR) technique successfully in spite of a monolayer structure.

CONCLUSION

Cell response for biomedical applications could be induced by designing smart materials that should elicit specific interactions with cell and thereby direct cell attachment, proliferation, differentiation and extracellular matrix production and organization. Interfaces between engineered materials and stem cells play a critical role in biomedical applications where the interaction between cells and material surface dictates cell performance and the success of an implanted device. Cells can act as mechano-sensitive units responding to the mechanical stimulation of the extracellular matrix through focal adhesions and changes in cytoskeletal organization. The interactions between stem cells and their environment are complex and not completely understood. In particular the relative importance of different physical parameters on the biocompatibility of biomaterials for tissue engineering is poorly understood. Hence this chapter discusses some of the most important aspects of poly(lactic acid) based nanocomposite scaffold. Although the field of scaffold fabrication is now well established and advancing at a fast rate, more progress remains to be made, especially in engineering interaction between nanostructures and PLA, providing scaffolds that can support thick tissue structures.

Enhanced understanding of the signals within the microenvironment that regulate cell fate has led to the development of increasingly sophisticated polymeric biomaterials for tissue engineering and regenerative medicine applications. This advancement is exemplified by biomaterials with precisely controlled scaffold architecture that regulate the spatio-temporal release of growth factors and morphogens, and respond dynamically to microenvironmental cues. PLA represents an excellent example of biopolymer for high performance scaffolds in tissue engineering. Further understanding of the biology, qualitatively and quantitatively, of cells within their microenvironments and at the tissue-material interface will expand the design space of future biomaterials.

ACKNOWLEDGMENTS

Authors wish to thank Prof. G. Gusmano, A. Bianco, Prof. S. Martino, Prof. A. Orlacchio, Dr. L. Visai, Prof. H. Arzate, Prof. F. Elisei, Prof. L. Latterini, Prof. P. Locci and Prof. E. Becchetti for fundamental collaborations in biological and material science fields.

REFERENCES

[1] Armentano, I., Dottori, M., Fortunati, E., Mattioli, S., and Kenny, J. M. *Polymer Degradation and Stability*, 95:2126–46 (2010).
[2] Gorrasi, G., Vittoria, V., Murariu, M., Ferreira, A. D. S., Alexandre, M., and Dubois, P. *Composite Science and Technology*, 9:627–32 (2008).
[3] Fortunati, E., D'Angelo, F., Martino, S., Orlacchio, A., Kenny, J. M., and Armentano, I. *Carbon*, 49:23709 (2011).
[4] Zhang, L. and Webster, T. J. *Nano Today*, 4:66–80 (2009).
[5] Chen, C., Lv, G., Pan, C., Song, M., Wu, C., Guo, D., et al. *Biomedical Materials*, 2:L1L4 (2007).
[6] Jee, K. S., Park, H. D., Park, K. D., Kim, Y. H., and Shin, J. W. *Biomacromolecules*, 5:187781 (2004).
[7] Armentano, I., Marinucci, L., Dottori, M., Balloni, S., Fortunati, E., Pennacchi, M., et al. *Journal of Biomaterials Science Polymer. Edition*, 22:541-556 (2010).
[8] Armentano, I., Ciapetti, G., Pennacchi, M., Dottori, M., Devescovi, V., Granchi, D., et al. *Journal of Applied Polymer Science*, 114:3602–11 (2009).
[9] Kim, H. W., Lee, H. H. and Chun, G. S. *Journal of Biomedical Material Research*, 85A:651–63 (2008).
[10] Prabhakaran, M. P., Venugopal, J. and Ramakrishna, S. *Acta Biomaterialia*, 5:288493 (2009).
[11] Bianco, A., Bozzo, B. M., Cacciotti, I., Del Gaudio, C., Armentano, I., Dottori, M., et al. *Journal of Bioactive and Compatible Polymers*, 26(3):225–41 (2011).
[12] Zhang, P., Hong, Z., Yub, T., Chen, X., and Jinga, X. *Biomaterials*, 30:58–70 (2009).
[13] Dresselhaus, M. S., Dresselhaus, G. and Eklund, P. C. *Science of Fullerenes and Carbon Nanotubes*. San Diego: Academic Press (1996).
[14] Dottori, M., Armentano, I., Fortunati, E., and Kenny, J. M. *Journal of Applied Polymer Science*,, 19:354452 (2011).
[15] Shaffer, M., Demchuk, V., Alstadt, V., Windle, A. H., Sandler, J., Werner, P. *Composites Part A*, 33:1033–1039 (2002).
[16] Walters, D. A., Ericson, L. M., Casavant, M. J., Liu, J., Colbert, D. T., Smith, K. A., and Smalley, R. E. *Applied Physics Letters*, 74:3803–3805 (1999).

[17] Mamedov, A. A., Kotov, N. A., Prato, M., Guldi, D. M., Wicksted, J. P., and Hirsch, A. *Nature Materials*, 1:190–194 (2002).

[18] Zhang, X., Liu, T., Sreekumar, T. V., Kumar, S., Moore, V. C., Hauge, R. H., and Smalley, R. E. *Nano Letters*, 3:1285–1288 (2003).

[19] Teo, K. B. K., Singh, C., Chhowalla, M., and Milne, W. I. *Catalytic synthesis of carbon nanotubes and nanofibers. Encyclopedia of Nanoscience and Nanotechnology*, volume X, 1-22. Edited by hs nalwa edition (2003).

[20] Jia, G., Wang, H., Yan, L., Wang, X., Pei, R., Yan, T., Zhao, Y., and Guo, X. *Environmental Science & Technology*, 39:1378–83 (2005).

[21] Harrison, B. S. and Atala, A. *Biomaterials*, 28:344–53 (2007).

[22] Tian, F., Cui, D., Schwarz, H., Estrada, G. G., and Kobayashi, H. *Toxicology In Vitro*, 20:1202–12 (2006).

[23] Osswald, S., Yushin, G., Mochalin, V., Kucheyev, S. O., and Gogotsi, Y. *Journal of the American Chemical Society.*, 128:11635–42 (2006).

[24] Decuzzi, P., Pasqualini, R., Arap, W., and Ferrari, M. *Pharmaceutical Research*, 26:235–43 (2009).

[25] He, Q., Yuan, W., Liu, J., and Zhang, R. *Journal Material Science: Material in Medicine*, 19:559–65 (2008).

[26] Xie, H. and Smith, J. W. *Journal of Nanobiotechnology*, 8:1–8 (2010).

[27] Faraji, A. H. and Wipf, P. *Bioorganic & Medical Chemistry*, 17:2950–62 (2009).

[28] Mora-Huertas, C. E., Fessi, H. and Elaissari, A. *Advance in Colloid and Interface*, 163:90–122 (2011).

[29] Mohanraj, V. J. and Chen, Y. *Tropical Journalof Pharmaceutical Research*, 5(1):561–73 (2006).

[30] Lassalle, V. and Ferreira, M. L. *Macromolecular Bioscience*, 7:767–783 (2007).

[31] Alexis, F., Pridgen, E., Molnar, L. K., and Farokhzad, O. C. *Molecular Pharmacology*, 5:505–15 (2008).

[32] Rezwan, K., Chen, Q. Z., Blaker, J. J., and Boccaccini, A. R. *Biomaterials*, 27:3413–3431 (2006).

[33] Klein, C. P. A. T., Driessen, A. A., De Groot, K., and Van den Hooff, A. *Journal of Biomedical Materials Research*, 17:769–784 (2004).

[34] Daculsi, G., Laboux, O., Malard, O., and Weiss, P. *Journal Material Science: Material in Medicine*,, 14:195–200 (2003).

[35] Wagoner Johnson, A. J. and Herschler, B. A. *Acta Biomaterialia*, 7:16–30 (2011).

[36] Rahaman, M. N., Day, D. E., Bal, B. S., Fu, Q., Jung, S. B., Bonewald, L. F., and Tomsia, A. P. *Acta Biomaterialia*, 7:2355–73 (2011).

[37] LeGeros, R. Z., Lin, S., Rohanizadeh, R., Mijares, D., and LeGeros, J. P. *Journal Material Science: Material in Medicine*,, 14:201–209 (2003).

[38] Durucan, C. and Brown, P. W. *Journal of Biomedical Materials Research*, 51:717–725 (2000).

[39] Bianco, A., Cacciotti, I., Lombardi, M., Montanaro, L., and Gusmano, G. *Journal of Therma. Analalysis and Calorimetry*, 88:237–243 (2007).

[40] Okuda, T., Ioku, K., Yonezawa, I., Minagi, H., Gonda, Y., Kawachi, G., et al. *Biomaterials*, 29:2719–2728 (2008).

[41] Peto, G., Molnar, G. L., Paszti, Z., Geszti, O., Beck, A., and Guczi, L. *Material Science & Engineering C*, 19:95–99 (2002).

[42] Panacek, A., Kvitek, L., Prucek, R., Kolar, M., Vecerova, R., Pizurova, N., Sharma, V. K., Nevecna, T., and Zboril, R. *Journal of Physical Chemistry B*, 110:16248–16253 (2006).

[43] Armentano, I., Fortunati, E., Latterini, L., Rinaldi, S., Saino, E., Visai, L., Elisei, F., and Kenny, J. M. *Journal of Nanostructured Polymers and Nanocomposites*, 6:110–118 (2010).

[44] Van der Zande, M., Jurado, M. J., Alava, J. I., Boerman, O. C., Walboomers, X. F., Olalde, B., and Jansen, J. A. *Journal of Tissue Engineering and Regenerative Medicine*, 5:476–482 (2011).

[45] Weigel, T., Schinkel, G. and Lendlein, A. *Expert Review of medical Devices*, 3:835–851 (2006).

[46] Hollister, S. J. *Nature Materials*, 7:518–24 (2005).

[47] Badami, A. S., Kreke, M. R., Thompson, M. S., Riffle, J. S., and Goldstein, A. S. *Biomaterials*, 27:596–606 (2006).

[48] He, L., Liao, S., Quan, D., Ma, K., Chan, C., Ramakrishna, S., et al. *Acta Biomaterialia*, 6:2960–9 (2010).

[49] Mikos, A. G., Thorsen, A. J., Gzerwonka, L. A., Bao, Y., Langer, R., Winslow, D. N., and Vacanti, J. P. *Polymer* 35: 1068-1077 (1994).

[50] Andriano, K. P., Tabata, Y., Ikada, Y., and Heller, J. *Journal of Biomedical Materials Research*,, 48:602–612 (1999).

[51] Park, A., Wu, B. and Griffith, L. G. *Journal of Biomaterials Science Polymer Edition*, 9:89–110 (1998).

[52] Mathieu, L. M., Mueller, T., Bourban, P. E., Pioletti, D. P., Mller, R., and Manson, J. A. E. *Biomaterials*, 27:905–16 (2005).

[53] Nam, Y. S., Yoon, J. J. and Park, T. G. *Journal of Biomedical Materials Research*,,, 53:1–7, (2000).

[54] Schugens, Ch., Maquet, V., Grandfils, Ch., Jerome, R., and Teyssie, Ph.. *Journal of Biomedical Materials Research*,,, 30:449–461 (1996).
[55] Yang, F., Murugan, R., Wang, S., and Ramakrishna, S. *Biomaterials*, 26:2603–10 (2005).
[56] Tan, S. H., Inai, R., Kotaki, M., and Ramakrishna, S. *Polymer*, 46:6128–34 (2005).
[57] Molamma P. Prabhakaran and Ramakrishna S. Venugopal J. *Acta Biomaterialia*, 5:2884–2893 (2009).
[58] Patel, S., Kurpinski, K., Quigley, R., Gao, H., Hsiao, B., Poo, M. M., et al. *Nano Letters*, 7:2122–8 (2007).
[59] D'Angelo, F., Armentano, I., Cacciotti, I., Quattrocelli, M., Del Gaudio, C., Fortunati, E., et al. *Biomacromolecules*. Accepted DOI: 10.1021/bm3000716
[60] Kim, H. W., Lee, H. H. and Knowles, J. C. *Journal of Biomedical Materials Research* A,, 79:643–649 (2006).
[61] Quirk, R. A., France, R. M., Shakesheff, K. M., and Howdle, S. M. *Current Opinion in Solid State and Materials Science*, 8:313321 (2004).
[62] Mathieu, L. M., Mueller, T. L., Bourbana, P. E., and Manson, J. A. E., Pioletti, D. P., Muller, R. *Biomaterials*, 27:905–916 (2006).
[63] Bhler, M., Bourban, P. E. and Manson, J. A. E. *Composite Science and Technology*, 68:820–8 (2008).
[64] Hu, Y., Zhang, C., Zhang, S., Xiong, Z., and Xu, J. *Journal of Biomedical Materials Research Part A*, 67:591–598 (2003).
[65] Martino, S., D'Angelo, F., Armentano, I., Kenny, J. M., and Orlacchio, A. *Biotechnology Advances*, 30:338–351 (2011).
[66] Kumari, A., Yadav, S. K. and Yadav, S. C. *Colloids and Surfaces B: Biointerfaces*, 75:1–18 (2010).
[67] Mei, F., Zhong, J., Yang, X., Ouyang, X., Zhang, S., Hu, X., et al. *Biomacromolecules*, 8:3729–35 (2007).
[68] Stone, D. A. and Korley, L. T. J. *Macromolecules*, 43:9217–26 (2010).
[69] Deng, X. M., Hao, J. Y. and Wang, C. S. *Biomaterials*, 22:2867–73 (2001).
[70] Wilberforce, S. I. J., Finlayson, C. E., Best, S. M., and Cameron, R. E. *Polymer,* 52: 2883-2890 (2011).
[71] Borum-Nicholas, L. and Wilson, Jr. O. C. *Biomaterials*, 24:3671-3679 (2003).
[72] Li, J., Lu, X. L. and Zheng, Y. F. *Appied.Surface Science*, 255:494–497 (2008).

[73] Sun, J. R., Deng, M. X., Chen, X. S., Jing, X. B., Hong, Z. K., Qiu, X. Y. *Polymer*, 45:6705–6713 (2004).

[74] Hong, Z., Zhang, P., He, C., Qiu, X., Liu, A., Chen Li, et al. *Biomaterials*, 26:6296–304 (2005).

[75] Oksman, K., Skrifvars, M. and Selin, J. F. *Composite Science & Technology*, 63:1317–24 (2003).

[76] Zhang, Q., Mochalin, V. N., Neitzel, I., Knoke, I. Y., Han, J., Klug, C. A., Zhou, J. G., Lelkes, P. I., and Gogotsi, Y. *Biomaterials*, 32:87–94 (2011).

[77] Wang, Y. L., Discher, D. E., Janmey, P. *Science*, 310:1139–43 (2005).

[78] Pittenger, M. F., Mackay, A. M., Beck, S. C., Jaiswal, R. K., Douglas, R., Mosca, J. D., Moorman, M. A., Simonetti, D. W., Craig, S., and Marshak, D. R. *Science*, 284:143–7 (1999).

[79] Lizundia, E., Sarasua, J.R., D'angelo, F., Orlacchio, A., Martino, S., Kenny, J. M., Armentano, I. *Macromolecular Bioscience Accepted.*

[80] Stauffer, D. and Aharony, A. *Introduction to Percolation Theory*. 1992.

[81] Kinloch, I. A., Shaffer, M. S. P., Sandler, J. K. W., Kirk, J. E., and Windle, A. H. *Polymer*, 44:58935899 (2003).

[82] Supronowicz, P. R., Ajayan, P. M., Ullmann, K. R., Arulanandam, B. P., Metzger, D. W., and Bizios, R. *Journal of Biomedical Materials Research*, 3:499–506 (2002).

[83] Sepahvand, R., Adeli, M., Astinchap, B., and Kabiri, R. *Journal of Nanoparticle Research*, 10:1309–1318 (2008).

INDEX

D

N

T